Particle Physics: A Comprehensive Introduction

Particle Physics: A Comprehensive Introduction

Edited by
Aidan Butler

Larsen & Keller
www.larsen-keller.com

Particle Physics: A Comprehensive Introduction
Edited by Aidan Butler
ISBN: 978-1-63549-214-9 (Hardback)

 Larsen & Keller

Published by Larsen and Keller Education,
5 Penn Plaza,
19th Floor,
New York, NY 10001, USA

Cataloging-in-Publication Data

Particle physics : a comprehensive introduction / edited by Aidan Butler.
    p. cm.
Includes bibliographical references and index.
ISBN 978-1-63549-214-9
1. Particles (Nuclear physics). 2. Particles. 3. Nuclear physics.
I. Butler, Aidan.
QC776 .P37 2017
539.72--dc23

The publisher's policy is to use permanent paper from mills that operate a sustainable forestry policy. Furthermore, the publisher ensures that the text paper and cover boards used have met acceptable environmental accreditation standards.

Printed and bound in the United States of America.

For more information regarding Larsen and Keller Education and its products, please visit the publisher's website www.larsen-keller.com

Table of Contents

Preface

As a branch of physics, particle physics concerns itself with the study of particles and their nature. It includes the study of the radiation and matter present in the particles. This field of study is concerned with subatomic particles and atomic constituents like quarks, neutrons, protons, electrons and baryons, etc. This book presents the complex subject of particle physics in the most comprehensible and easy to understand language. The various sub-fields of the subject along with technological progress that have future implications are glanced at in it. The topics covered in this extensive text deal with the core subjects of this area. This textbook will serve as a reference to a broad spectrum of readers.

A short introduction to every chapter is written below to provide an overview of the content of the book:

Chapter 1 - The branch of physics that deals with the nature of particles that make up matter is known as particle physics. Particle physics studies the smallest detectable particles. This chapter is an overview of the subject matter incorporating all the major aspects of particle physics; **Chapter 2** - The fundamentals of particle physics elucidated in this text are field, weak interaction, strong interaction and electroweak interaction. A field in physics has physical quantity; it has value for each point. The two out of the four fundamental interactions that occur in particle physics are strong interactions and weak interactions. The text helps the reader in developing an in- depth understanding of the fundamentals of particle physics; **Chapter 3** - The theories of particle physics discussed in this chapter are quantum chromodynamics, quantum field theory, gauge theory and peccei-quinn theory. Quantum chromodynamics is the theory of strong interactions whereas quantum field theory framework is a part of particle physics that is used for constructing quantum mechanical models of subatomic particles. The aim of this section is to explore the theories of particle physics; **Chapter 4** - The theory of particle physics that concerns with the electromagnetic, weak and strong nuclear interactions is known as standard model of particle physics. The topics discusses in this chapter are Higgs mechanism, Cabibbo-Kobayashi-Maskawa matrix and spontaneous symmetry breaking. The major components of conceptual models are discussed in this chapter; **Chapter 5** - The elementary particles elucidated in the section are elementary particle, quark, photon, lepton, gluon, Higgs boson, fermion, antimatter and boson. Elementary particle is the particle whose structure is unknown and quark is the elementary particle of matter. Quark is also one of the central constitutes of matter. This text is an overview of the subject matter incorporating all the major aspects of elementary particles of particle physics; **Chapter 6** - Particle physics has a number of technologies associated with it; some of these are particle accelerator, accelerator physics and particle physics experiments. Particle accelerators are machines that use electromagnetic fields to boost charged particles in order to include them in well-defined beams. This chapter elucidates the main technologies related to particle physics; **Chapter 7 -** The idea of particles and matter has existed in natural philosophy since 6th century BC. Particles have been discovered and researched; these particles are studied under high energies. This text helps the reader in developing an understanding related to the history of particle physics.

I extend my sincere thanks to the publisher for considering me worthy of this task. Finally, I thank my family for being a source of support and help.

Editor

Introduction to Particle Physics

The branch of physics that deals with the nature of particles that make up matter is known as particle physics. Particle physics studies the smallest detectable particles. This chapter is an overview of the subject matter incorporating all the major aspects of particle physics.

Particle physics (also high energy physics) is the branch of physics that studies the nature of the particles that constitute *matter* (particles with mass) and *radiation* (massless particles). Although the word "particle" can refer to various types of very small objects (e.g. protons, gas particles, or even household dust), "particle physics" usually investigates the irreducibly smallest detectable particles and the irreducibly fundamental force fields necessary to explain them. By our current understanding, these elementary particles are excitations of the quantum fields that also govern their interactions. The currently dominant theory explaining these fundamental particles and fields, along with their dynamics, is called the Standard Model. Thus, modern particle physics generally investigates the Standard Model and its various possible extensions, e.g. to the newest "known" particle, the Higgs boson, or even to the oldest known force field, gravity.

Subatomic Particles

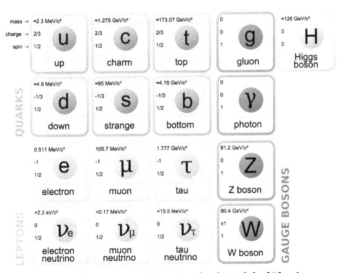

The particle content of the Standard Model of Physics

Modern particle physics research is focused on subatomic particles, including atomic

constituents such as electrons, protons, and neutrons (protons and neutrons are composite particles called baryons, made of quarks), produced by radioactive and scattering processes, such as photons, neutrinos, and muons, as well as a wide range of exotic
particles. Dynamics of particles is also governed by quantum mechanics; they exhibit
wave–particle duality, displaying particle-like behaviour under certain experimental
conditions and wave-like behaviour in others. In more technical terms, they are described by quantum state vectors in a Hilbert space, which is also treated in quantum
field theory. Following the convention of particle physicists, the term *elementary particles* is applied to those particles that are, according to current understanding, presumed to be indivisible and not composed of other particles.

Elementary Particles					
	Types	**Generations**	**Antiparticle**	**Colors**	**Total**
Quarks	2	3	Pair	3	36
Leptons			Pair	None	12
Gluons	1	1	Own	8	8
Photon			Own	None	1
Z Boson			Own		1
W Boson			Pair		2
Higgs			Own		1
Total number of (known) elementary particles:					61

All particles and their interactions observed to date can be described almost entirely by
a quantum field theory called the Standard Model. The Standard Model, as currently
formulated, has 61 elementary particles. Those elementary particles can combine to
form composite particles, accounting for the hundreds of other species of particles that
have been discovered since the 1960s. The Standard Model has been found to agree
with almost all the experimental tests conducted to date. However, most particle physicists believe that it is an incomplete description of nature and that a more fundamental theory awaits discovery. In recent years, measurements of neutrino mass have
provided the first experimental deviations from the Standard Model.

History

The idea that all matter is composed of elementary particles dates to at least the 6th
century BC. In the 19th century, John Dalton, through his work on stoichiometry, concluded that each element of nature was composed of a single, unique type of particle.
The word *atom*, after the Greek word *atomos* meaning "indivisible", denotes the smallest particle of a chemical element since then, but physicists soon discovered that *atoms*
are not, in fact, the fundamental particles of nature, but conglomerates of even smaller
particles, such as the electron. The early 20th-century explorations of nuclear physics

and quantum physics culminated in proofs of nuclear fission in 1939 by Lise Meitner (based on experiments by Otto Hahn), and nuclear fusion by Hans Bethe in that same year; both discoveries also led to the development of nuclear weapons. Throughout the 1950s and 1960s, a bewildering variety of particles were found in scattering experiments. It was referred to as the "particle zoo". That term was deprecated after the formulation of the Standard Model during the 1970s in which the large number of particles was explained as combinations of a (relatively) small number of fundamental particles.

Standard Model

The current state of the classification of all elementary particles is explained by the Standard Model. It describes the strong, weak, and electromagnetic fundamental interactions, using mediating gauge bosons. The species of gauge bosons are the gluons, W−, W+ and Z bosons, and the photons. The Standard Model also contains 24 fundamental particles, (12 particles and their associated anti-particles), which are the constituents of all matter. Finally, the Standard Model also predicted the existence of a type of boson known as the Higgs boson. Early in the morning on 4 July 2012, physicists with the Large Hadron Collider at CERN announced they had found a new particle that behaves similarly to what is expected from the Higgs boson.

Experimental Laboratories

In particle physics, the major international laboratories are located at the:

- Brookhaven National Laboratory (Long Island, United States). Its main facility is the Relativistic Heavy Ion Collider (RHIC), which collides heavy ions such as gold ions and polarized protons. It is the world's first heavy ion collider, and the world's only polarized proton collider.

- Budker Institute of Nuclear Physics (Novosibirsk, Russia). Its main projects are now the electron-positron colliders VEPP-2000, operated since 2006, and VEPP-4, started experiments in 1994. Earlier facilities include the first electron-electron beam-beam collider VEP-1, which conducted experiments from 1964 to 1968; the electron-positron colliders VEPP-2, operated from 1965 to 1974; and, its successor VEPP-2M, performed experiments from 1974 to 2000.

- CERN, (Conseil Européen pour la Recherche Nucléaire) (Franco-Swiss border, near Geneva). Its main project is now the Large Hadron Collider (LHC), which had its first beam circulation on 10 September 2008, and is now the world's most energetic collider of protons. It also became the most energetic collider of heavy ions after it began colliding lead ions. Earlier facilities include the Large Electron–Positron Collider (LEP), which was stopped on 2 November 2000 and then dismantled to give way for LHC; and the Super Proton Synchrotron, which is being reused as a pre-accelerator for the LHC.

- DESY (Deutsches Elektronen-Synchrotron) (Hamburg, Germany). Its main facility is the Hadron Elektron Ring Anlage (HERA), which collides electrons and positrons with protons.

- Fermi National Accelerator Laboratory (Fermilab), (Batavia, United States). Its main facility until 2011 was the Tevatron, which collided protons and antiprotons and was the highest-energy particle collider on earth until the Large Hadron Collider surpassed it on 29 November 2009.

- Institute of High Energy Physics (IHEP), (Beijing, China). IHEP manages a number of China's major particle physics facilities, including the Beijing Electron Positron Collider (BEPC), the Beijing Spectrometer (BES), the Beijing Synchrotron Radiation Facility (BSRF), the International Cosmic-Ray Observatory at Yangbajing in Tibet, the Daya Bay Reactor Neutrino Experiment, the China Spallation Neutron Source, the Hard X-ray Modulation Telescope (HXMT), and the Accelerator-driven Sub-critical System (ADS) as well as the Jiangmen Underground Neutrino Observatory (JUNO).

- KEK, (Tsukuba, Japan). It is the home of a number of experiments such as the K2K experiment, a neutrino oscillation experiment and Belle, an experiment measuring the CP violation of B mesons.

- SLAC National Accelerator Laboratory, (Menlo Park, United States). Its 2-mile-long linear particle accelerator began operating in 1962 and was the basis for numerous electron and positron collision experiments until 2008. Since then the linear accelerator is being used for the Linac Coherent Light Source X-ray laser as well as advanced accelerator design research. SLAC staff continue to participate in developing and building many particle physics experiments around the world.

Many other particle accelerators do exist.

The techniques required to do modern, experimental, particle physics are quite varied and complex, constituting a sub-specialty nearly completely distinct from the theoretical side of the field.

Theory

Theoretical particle physics attempts to develop the models, theoretical framework, and mathematical tools to understand current experiments and make predictions for future experiments. There are several major interrelated efforts being made in theoretical particle physics today. One important branch attempts to better understand the Standard Model and its tests. By extracting the parameters of the

Standard Model, from experiments with less uncertainty, this work probes the limits of the Standard Model and therefore expands our understanding of nature's building blocks. Those efforts are made challenging by the difficulty of calculating quantities in quantum chromodynamics. Some theorists working in this area refer to themselves as phenomenologists and they may use the tools of quantum field theory and effective field theory. Others make use of lattice field theory and call themselves *lattice theorists*.

Another major effort is in model building where model builders develop ideas for what physics may lie beyond the Standard Model (at higher energies or smaller distances). This work is often motivated by the hierarchy problem and is constrained by existing experimental data. It may involve work on supersymmetry, alternatives to the Higgs mechanism, extra spatial dimensions (such as the Randall-Sundrum models), Preon theory, combinations of these, or other ideas.

A third major effort in theoretical particle physics is string theory. *String theorists* attempt to construct a unified description of quantum mechanics and general relativity by building a theory based on small strings, and branes rather than particles. If the theory is successful, it may be considered a "Theory of Everything", or "TOE".

There are also other areas of work in theoretical particle physics ranging from particle cosmology to loop quantum gravity.

This division of efforts in particle physics is reflected in the names of categories on the arXiv, a preprint archive: hep-th (theory), hep-ph (phenomenology), hep-ex (experiments), hep-lat (lattice gauge theory).

Practical Applications

In principle, all physics (and practical applications developed there from) can be derived from the study of fundamental particles. In practice, even if "particle physics" is taken to mean only "high-energy atom smashers", many technologies have been developed during these pioneering investigations that later find wide uses in society. Cyclotrons are used to produce medical isotopes for research and treatment (for example, isotopes used in PET imaging), or used directly for certain cancer treatments. The development of Superconductors has been pushed forward by their use in particle physics. The World Wide Web and touchscreen technology were initially developed at CERN.

Additional applications are found in medicine, national security, industry, computing, science, and workforce development, illustrating a long and growing list of beneficial practical applications with contributions from particle physics.

Future

The primary goal, which is pursued in several distinct ways, is to find and understand what physics may lie beyond the standard model. There are several powerful experimental reasons to expect new physics, including dark matter and neutrino mass. There are also theoretical hints that this new physics should be found at accessible energy scales.

Much of the effort to find this new physics are focused on new collider experiments. The Large Hadron Collider (LHC) was completed in 2008 to help continue the search for the Higgs boson, supersymmetric particles, and other new physics. An intermediate goal is the construction of the International Linear Collider (ILC), which will complement the LHC by allowing more precise measurements of the properties of newly found particles. In August 2004, a decision for the technology of the ILC was taken but the site has still to be agreed upon.

In addition, there are important non-collider experiments that also attempt to find and understand physics beyond the Standard Model. One important non-collider effort is the determination of the neutrino masses, since these masses may arise from neutrinos mixing with very heavy particles. In addition, cosmological observations provide many useful constraints on the dark matter, although it may be impossible to determine the exact nature of the dark matter without the colliders. Finally, lower bounds on the very long lifetime of the proton put constraints on Grand Unified Theories at energy scales much higher than collider experiments will be able to probe any time soon.

In May 2014, the Particle Physics Project Prioritization Panel released its report on particle physics funding priorities for the United States over the next decade. This report emphasized continued U.S. participation in the LHC and ILC, and expansion of the Long Baseline Neutrino Experiment, among other recommendations.

References

- Braibant, S.; Giacomelli, G.; Spurio, M. (2009). Particles and Fundamental Interactions: An Introduction to Particle Physics. Springer. pp. 313–314. ISBN 978-94-007-2463-1.

- Mann, Adam (28 March 2013). "Newly Discovered Particle Appears to Be Long-Awaited Higgs Boson - Wired Science". Wired.com. Retrieved 6 February 2014.

- "Particle Physics and Astrophysics Research". The Henryk Niewodniczanski Institute of Nuclear Physics. Retrieved 31 May 2012.

2

Fundamentals of Particle Physics

The fundamentals of particle physics elucidated in this text are field, weak interaction, strong interaction and electroweak interaction. A field in physics has physical quantity; it has value for each point. The two out of the four fundamental interactions that occur in particle physics are strong interactions and weak interactions. The text helps the reader in developing an in- depth understanding of the fundamentals of particle physics.

Field (Physics)

In physics, a field is a physical quantity that has a value for each point in space and time. For example, on a weather map, the surface wind velocity is described by assigning a vector to each point on a map. Each vector represents the speed and direction of the movement of air at that point. As another example, an electric field can be thought of as a "condition in space" emanating from an electric charge and extending throughout the whole of space. When a test electric charge is placed in this electric field, the particle accelerates due to a force. Physicists have found the notion of a field to be of such practical utility for the analysis of forces that they have come to think of a force as due to a field.

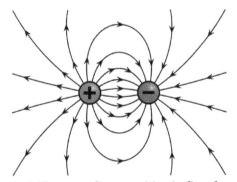

Illustration of the electric field surrounding a positive (red) and a negative (blue) charge.

In the modern framework of the quantum theory of fields, even without referring to a test particle, a field occupies space, contains energy, and its presence eliminates a true vacuum. This led physicists to consider electromagnetic fields to be a physical entity, making the field concept a supporting paradigm of the edifice of modern physics. "The fact that the electromagnetic field can possess momentum and energy makes it very real... a particle makes a field, and a field acts on another particle, and the field has such familiar properties as energy content and momentum, just as particles can have".

In practice, the strength of most fields has been found to diminish with distance to the point of being undetectable. For instance the strength of many relevant classical fields, such as the gravitational field in Newton's theory of gravity or the electrostatic field in classical electromagnetism, is inversely proportional to the square of the distance from the source (i.e. they follow the Gauss's law). One consequence is that the Earth's gravitational field quickly becomes undetectable on cosmic scales.

A field can be classified as a scalar field, a vector field, a spinor field or a tensor field according to whether the represented physical quantity is a scalar, a vector, a spinor or a tensor, respectively. A field has a unique tensorial character in every point where it is defined: i.e. a field cannot be a scalar field somewhere and a vector field somewhere else. For example, the Newtonian gravitational field is a vector field: specifying its value at a point in spacetime requires three numbers, the components of the gravitational field vector at that point. Moreover, within each category (scalar, vector, tensor), a field can be either a *classical field* or a *quantum field*, depending on whether it is characterized by numbers or quantum operators respectively. In fact in this theory an equivalent representation of field is a field particle, namely a boson.

History

To Isaac Newton his law of universal gravitation simply expressed the gravitational force that acted between any pair of massive objects. When looking at the motion of many bodies all interacting with each other, such as the planets in the Solar System, dealing with the force between each pair of bodies separately rapidly becomes computationally inconvenient. In the eighteenth century, a new quantity was devised to simplify the bookkeeping of all these gravitational forces. This quantity, the gravitational field, gave at each point in space the total gravitational force which would be felt by an object with unit mass at that point. This did not change the physics in any way: it did not matter if you calculated all the gravitational forces on an object individually and then added them together, or if you first added all the contributions together as a gravitational field and then applied it to an object.

The development of the independent concept of a field truly began in the nineteenth century with the development of the theory of electromagnetism. In the early stages, André-Marie Ampère and Charles-Augustin de Coulomb could manage with Newton-style laws that expressed the forces between pairs of electric charges or electric currents. However, it became much more natural to take the field approach and express these laws in terms of electric and magnetic fields; in 1849 Michael Faraday became the first to coin the term "field".

The independent nature of the field became more apparent with James Clerk Maxwell's discovery that waves in these fields propagated at a finite speed. Consequently, the forces on charges and currents no longer just depended on the positions and velocities of other charges and currents at the same time, but also on their positions and velocities in the past.

Maxwell, at first, did not adopt the modern concept of a field as fundamental quantity that could independently exist. Instead, he supposed that the electromagnetic field expressed the deformation of some underlying medium—the luminiferous aether—much like the tension in a rubber membrane. If that were the case, the observed velocity of the electromagnetic waves should depend upon the velocity of the observer with respect to the aether. Despite much effort, no experimental evidence of such an effect was ever found; the situation was resolved by the introduction of the special theory of relativity by Albert Einstein in 1905. This theory changed the way the viewpoints of moving observers should be related to each other in such a way that velocity of electromagnetic waves in Maxwell's theory would be the same for all observers. By doing away with the need for a background medium, this development opened the way for physicists to start thinking about fields as truly independent entities.

In the late 1920s, the new rules of quantum mechanics were first applied to the electromagnetic fields. In 1927, Paul Dirac used quantum fields to successfully explain how the decay of an atom to lower quantum state lead to the spontaneous emission of a photon, the quantum of the electromagnetic field. This was soon followed by the realization (following the work of Pascual Jordan, Eugene Wigner, Werner Heisenberg, and Wolfgang Pauli) that all particles, including electrons and protons, could be understood as the quanta of some quantum field, elevating fields to the status of the most fundamental objects in nature. That said, John Wheeler and Richard Feynman seriously considered Newton's pre-field concept of action at a distance (although they set it aside because of the ongoing utility of the field concept for research in general relativity and quantum electrodynamics).

Classical Fields

There are several examples of classical fields. Classical field theories remain useful wherever quantum properties do not arise, and can be active areas of research. Elasticity of materials, fluid dynamics and Maxwell's equations are cases in point.

Some of the simplest physical fields are vector force fields. Historically, the first time that fields were taken seriously was with Faraday's lines of force when describing the electric field. The gravitational field was then similarly described.

Newtonian Gravitation

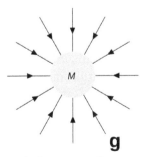

In classical gravitation, mass is the source of an attractive gravitational field g.

A classical field theory describing gravity is Newtonian gravitation, which describes the gravitational force as a mutual interaction between two masses.

Any body with mass M is associated with a gravitational field g which describes its influence on other bodies with mass. The gravitational field of M at a point r in space corresponds to the ratio between force F that M exerts on a small or negligible test mass m located at r and the test mass itself:

$$\mathbf{g}(\mathbf{r}) = \frac{\mathbf{F}(\mathbf{r})}{m}.$$

Stipulating that m is much smaller than M ensures that the presence of m has a negligible influence on the behavior of M.

According to Newton's law of universal gravitation, F(r) is given by

$$\mathbf{F}(\mathbf{r}) = -\frac{GMm}{r^2}\hat{\mathbf{r}},$$

where $\hat{\mathbf{r}}$ is a unit vector lying along the line joining M and m and pointing from m to M. Therefore, the gravitational field of M is

$$(\mathbf{r}) = \frac{\mathbf{F}(\mathbf{r})}{m} = -\frac{GM}{r^2}\hat{\mathbf{r}}.$$

The experimental observation that inertial mass and gravitational mass are equal to an unprecedented level of accuracy leads to the identity that gravitational field strength is identical to the acceleration experienced by a particle. This is the starting point of the equivalence principle, which leads to general relativity.

Because the gravitational force F is conservative, the gravitational field g can be rewritten in terms of the gradient of a scalar function, the gravitational potential $\Phi(r)$:

$$\mathbf{g}(\mathbf{r}) = -\nabla\Phi(\mathbf{r}).$$

Electromagnetism

Michael Faraday first realized the importance of a field as a physical quantity, during his investigations into magnetism. He realized that electric and magnetic fields are not only fields of force which dictate the motion of particles, but also have an independent physical reality because they carry energy.

These ideas eventually led to the creation, by James Clerk Maxwell, of the first unified field theory in physics with the introduction of equations for the electromagnetic field. The modern version of these equations is called Maxwell's equations.

Electrostatics

A charged test particle with charge q experiences a force F based solely on its charge. We can similarly describe the electric field E so that F = qE. Using this and Coulomb's law tells us that the electric field due to a single charged particle as

$$\mathbf{E} = \frac{1}{4\pi\epsilon_0}\frac{q}{r^2}\hat{\mathbf{r}}.$$

The electric field is conservative, and hence can be described by a scalar potential, $V(\mathbf{r})$:

$$\mathbf{E}(\mathbf{r}) = -\nabla V(\mathbf{r}).$$

Magnetostatics

A steady current I flowing along a path ℓ will exert a force on nearby moving charged particles that is quantitatively different from the electric field force described above. The force exerted by I on a nearby charge q with velocity v is

$$(\mathbf{r}) = q\mathbf{v}\times\mathbf{B}(\mathbf{r}),$$

where B(r) is the magnetic field, which is determined from I by the Biot–Savart law:

$$\mathbf{B}(\mathbf{r}) = \frac{\mu_0 I}{4\pi}\int\frac{d\ell\times d\hat{\mathbf{r}}}{r^2}.$$

The magnetic field is not conservative in general, and hence cannot usually be written in terms of a scalar potential. However, it can be written in terms of a vector potential, A(r):

$$\mathbf{B}(\mathbf{r}) = \nabla\times\mathbf{A}(\mathbf{r})$$

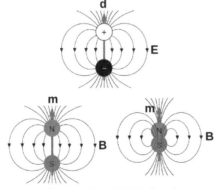

The E fields and B fields due to electric charges (black/white) and magnetic poles (red/blue). Top: E field due to an electric dipole moment d. Bottom left: B field due to a *mathematical* magnetic dipole m formed by two magnetic monopoles. Bottom right: B field due to a pure magnetic dipole moment m found in ordinary matter (*not* from monopoles).

Electrodynamics

In general, in the presence of both a charge density $\rho(r, t)$ and current density $J(r, t)$, there will be both an electric and a magnetic field, and both will vary in time. They are determined by Maxwell's equations, a set of differential equations which directly relate E and B to ρ and J.

Alternatively, one can describe the system in terms of its scalar and vector potentials V and A. A set of integral equations known as *retarded potentials* allow one to calculate V and A from ρ and J, and from there the electric and magnetic fields are determined via the relations

$$\mathbf{E} = -\nabla V - \frac{\partial \mathbf{A}}{\partial t}$$

$$\mathbf{B} = \nabla \times \mathbf{A}.$$

At the end of the 19th century, the electromagnetic field was understood as a collection of two vector fields in space. Nowadays, one recognizes this as a single antisymmetric 2nd-rank tensor field in spacetime.

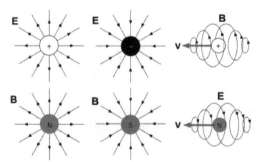

The E fields and B fields due to electric charges (black/white) and magnetic poles (red/blue). E fields due to stationary electric charges and B fields due to stationary magnetic charges (note in nature N and S monopoles do not exist). In motion (velocity v), an *electric* charge induces a B field while a *magnetic* charge (not found in nature) would induce an E field. Conventional current is used.

Gravitation in General Relativity

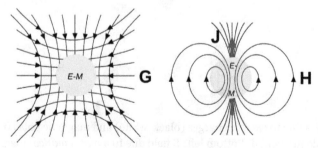

In general relativity, mass-energy warps space time (Einstein tensor G), and rotating asymmetric mass-energy distributions with angular momentum J generate GEM fields H

Einstein's theory of gravity, called general relativity, is another example of a field theory. Here the principal field is the metric tensor, a symmetric 2nd-rank tensor field in spacetime. This replaces Newton's law of universal gravitation.

Waves as Fields

Waves can be constructed as physical fields, due to their finite propagation speed and causal nature when a simplified physical model of an isolated closed system is set . They are also subject to the inverse-square law.

For electromagnetic waves, there are optical fields, and terms such as near- and far-field limits for diffraction. In practice, though the field theories of optics are superseded by the electromagnetic field theory of Maxwell.

Quantum Fields

It is now believed that quantum mechanics should underlie all physical phenomena, so that a classical field theory should, at least in principle, permit a recasting in quantum mechanical terms; success yields the corresponding quantum field theory. For example, quantizing classical electrodynamics gives quantum electrodynamics. Quantum electrodynamics is arguably the most successful scientific theory; experimental data confirm its predictions to a higher precision (to more significant digits) than any other theory. The two other fundamental quantum field theories are quantum chromodynamics and the electroweak theory.

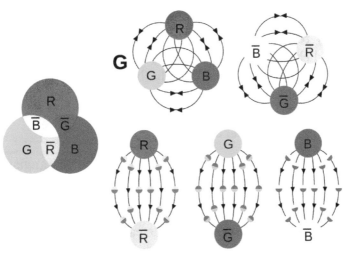

Fields due to color charges, like in quarks (G is the gluon field strength tensor). These are "colorless" combinations. Top: Color charge has "ternary neutral states" as well as binary neutrality (analogous to electric charge). Bottom: The quark/antiquark combinations.

In quantum chromodynamics, the color field lines are coupled at short distances by gluons, which are polarized by the field and line up with it. This effect increases within a short distance (around 1 fm from the vicinity of the quarks) making the color force increase within a short distance, confining the quarks within hadrons. As the field lines are pulled together tightly by gluons, they do not "bow" outwards as much as an electric field between electric charges.

These three quantum field theories can all be derived as special cases of the so-called standard model of particle physics. General relativity, the Einsteinian field theory of gravity, has yet to be successfully quantized. However an extension, thermal field theory, deals with quantum field theory at *finite temperatures*, something seldom considered in quantum field theory.

In BRST theory one deals with odd fields, e.g. Faddeev–Popov ghosts. There are different descriptions of odd classical fields both on graded manifolds and supermanifolds.

As above with classical fields, it is possible to approach their quantum counterparts from a purely mathematical view using similar techniques as before. The equations governing the quantum fields are in fact PDEs (specifically, relativistic wave equations (RWEs)). Thus one can speak of Yang–Mills, Dirac, Klein–Gordon and Schrödinger fields as being solutions to their respective equations. A possible problem is that these RWEs can deal with complicated mathematical objects with exotic algebraic properties (e.g. spinors are not tensors, so may need calculus over spinor fields), but these in theory can still be subjected to analytical methods given appropriate mathematical generalization.

Field Theory

Field theory usually refers to a construction of the dynamics of a field, i.e. a specification of how a field changes with time or with respect to other independent physical variables on which the field depends. Usually this is done by writing a Lagrangian or a Hamiltonian of the field, and treating it as the classical mechanics (or quantum mechanics) of a system with an infinite number of degrees of freedom. The resulting field theories are referred to as classical or quantum field theories.

The dynamics of a classical field are usually specified by the Lagrangian density in terms of the field components; the dynamics can be obtained by using the action principle.

It is possible to construct simple fields without any a priori knowledge of physics using only mathematics from several variable calculus, potential theory and partial differential equations (PDEs). For example, scalar PDEs might consider quantities such as amplitude, density and pressure fields for the wave equation and fluid dynamics; temperature/concentration fields for the heat/diffusion equations. Outside of physics proper (e.g., radiometry and computer graphics), there are even light fields. All these previous examples are scalar fields. Similarly for vectors, there are vector PDEs for displacement, velocity and vorticity fields in (applied mathematical) fluid dynamics, but vector calculus may now be needed in addition, being calculus over vector fields (as are these three quantities, and those for vector PDEs in general). More generally problems in continuum mechanics may involve for example, directional elasticity (from which comes the term *tensor*, derived from the Latin word for stretch), complex fluid flows or anisotropic diffusion, which are framed as matrix-tensor PDEs, and then require matrices or tensor fields, hence matrix or tensor calculus. It should be noted that the scalars (and hence the vectors, matrices and tensors) can be real or complex as both are fields in the abstract-algebraic/ring-theoretic sense.

In a general setting, classical fields are described by sections of fiber bundles and their dynamics is formulated in the terms of jet manifolds (covariant classical field theory).

In modern physics, the most often studied fields are those that model the four fundamental forces which one day may lead to the Unified Field Theory.

Symmetries of Fields

A convenient way of classifying a field (classical or quantum) is by the symmetries it possesses. Physical symmetries are usually of two types:

Spacetime Symmetries

Fields are often classified by their behaviour under transformations of spacetime. The terms used in this classification are:

- scalar fields (such as temperature) whose values are given by a single variable at each point of space. This value does not change under transformations of space.

- vector fields (such as the magnitude and direction of the force at each point in a magnetic field) which are specified by attaching a vector to each point of space. The components of this vector transform between themselves contravariantly under rotations in space. Similarly, a dual (or co-) vector field attaches a dual vector to each point of space, and the components of each dual vector transform covariantly.

- tensor fields, (such as the stress tensor of a crystal) specified by a tensor at each point of space. Under rotations in space, the components of the tensor transform in a more general way which depends on the number of covariant indices and contravariant indices.

- spinor fields (such as the Dirac spinor) arise in quantum field theory to describe particles with spin which transform like vectors except for the one of their component; in other words, when one rotates a vector field 360 degrees around a specific axis, the vector field turns to itself; however, spinors in same case turn to their negatives.

Internal Symmetries

Fields may have internal symmetries in addition to spacetime symmetries. For example, in many situations one needs fields which are a list of space-time scalars: $(\varphi_1, \varphi_2, \ldots \varphi_N)$. For example, in weather prediction these may be temperature, pressure, humidity, etc. In particle physics, the color symmetry of the interaction of quarks is an example of an internal symmetry of the strong interaction, as is the isospin or flavour symmetry.

If there is a symmetry of the problem, not involving spacetime, under which these components transform into each other, then this set of symmetries is called an *internal symmetry*. One may also make a classification of the charges of the fields under internal symmetries.

Statistical Field Theory

Statistical field theory attempts to extend the field-theoretic paradigm toward many-body systems and statistical mechanics. As above, it can be approached by the usual infinite number of degrees of freedom argument.

Much like statistical mechanics has some overlap between quantum and classical mechanics, statistical field theory has links to both quantum and classical field theories, especially the former with which it shares many methods. One important example is mean field theory.

Continuous Random Fields

Classical fields as above, such as the electromagnetic field, are usually infinitely differentiable functions, but they are in any case almost always twice differentiable. In contrast, generalized functions are not continuous. When dealing carefully with classical fields at finite temperature, the mathematical methods of continuous random fields are used, because thermally fluctuating classical fields are nowhere differentiable. Random fields are indexed sets of random variables; a continuous random field is a random field that has a set of functions as its index set. In particular, it is often mathematically convenient to take a continuous random field to have a Schwartz space of functions as its index set, in which case the continuous random field is a tempered distribution.

We can think about a continuous random field, in a (very) rough way, as an ordinary function that is almost everywhere, but such that when we take a weighted average of all the infinities over any finite region, we get a finite result. The infinities are not well-defined; but the finite values can be associated with the functions used as the weight functions to get the finite values, and that can be well-defined. We can define a continuous random field well enough as a linear map from a space of functions into the real numbers.

Weak Interaction

In particle physics, the weak interaction, the weak force or weak nuclear force, is one of the four known fundamental interactions of nature, alongside the strong interaction, electromagnetism, and gravitation. The weak interaction is responsible for radioactive decay, which plays an essential role in nuclear fission. The theory of the weak interaction is sometimes called quantum flavordynamics (QFD), in analogy with the terms QCD and QED, but the term is rarely used because the weak force is best understood in terms of electro-weak theory (EWT).

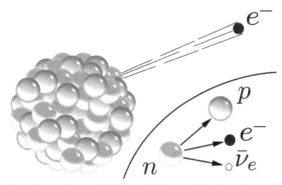

The radioactive beta decay is possibly due to the weak interaction, which transforms a neutron into: a proton, an electron, and an electron antineutrino.

In the Standard Model of particle physics, the weak interaction is caused by the emission or absorption of the force carriers, the W and Z bosons. All known fermions interact through the weak interaction. Fermions are particles that have half-integer spin. Spin is one of the fundamental properties of particles. A fermion can be an elementary particle, such as the electron, or it can be a composite particle, such as the proton. The masses of W^+, W^-, and Z bosons are each far greater than that of interacting protons or neutrons, which is consistent with the short range of the weak force. The force is termed *weak* because its field strength over a given distance is typically several orders of magnitude less than that of the strong nuclear force and electromagnetic force.

During the quark epoch of the early universe, the electroweak force separated into the electromagnetic and weak forces. Important examples of the weak interaction include beta decay, and the fusion of hydrogen into deuterium that powers the Sun's thermonuclear process. Most fermions will decay by a weak interaction over time. Such decay makes radiocarbon dating possible, as carbon-14 decays through the weak interaction to nitrogen-14. It can also create radioluminescence, commonly used in tritium illumination, and in the related field of betavoltaics.

Quarks, which make up composite particles like neutrons and protons, come in six "flavours" – up, down, strange, charm, top and bottom – which give those composite particles their properties. The weak interaction is unique in that it allows for quarks to swap their flavour for another. The swapping of those properties is mediated by the force carrier bosons. For example, during beta minus decay, a down quark within a neutron is changed into an up quark, converting the neutron to a proton and resulting in the emission of an electron and an electron antineutrino. Also, the weak interaction is the only fundamental interaction that breaks parity-symmetry, and similarly, the only one to break charge parity symmetry.

History

In 1933, Enrico Fermi proposed the first theory of the weak interaction, known as Fermi's interaction. He suggested that beta decay could be explained by a four-fermion interaction, involving a contact force with no range.

However, it is better described as a non-contact force field having a finite range, albeit very short. In 1968, Sheldon Glashow, Abdus Salam and Steven Weinberg unified the electromagnetic force and the weak interaction by showing them to be two aspects of a single force, now termed the electro-weak force.

The existence of the W and Z bosons was not directly confirmed until 1983.

Properties

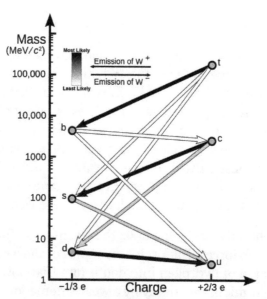

A diagram depicting the various decay routes due to the weak interaction and some indication of their likelihood. The intensity of the lines are given by the CKM parameters.

The weak interaction is unique in a number of respects:

- It is the only interaction capable of changing the flavor of quarks (i.e., of changing one type of quark into another).

- It is the only interaction that violates P or parity-symmetry. It is also the only one that violates CP symmetry.

- It is propagated by force carrier particles that have significant masses, an unusual feature which is explained in the Standard Model by the Higgs mechanism.

Due to their large mass (approximately 90 GeV/c²) these carrier particles, termed the W and Z bosons, are short-lived with a lifetime of under 10^{-24} seconds. The weak interaction has a coupling constant (an indicator of interaction strength) of between 10^{-7} and 10^{-6}, compared to the strong interaction's coupling constant of 1 and the electromagnetic coupling constant of about 10^{-2}; consequently the weak interaction is weak in terms of strength. The weak interaction has a very short range (around 10^{-17} to 10^{-16} m). At distances around 10^{-18} meters, the weak interaction has a strength of a similar magnitude to the electromagnetic force, but this starts to decrease exponentially with increasing distance. At distances of around 3×10^{-17} m, the weak interaction is 10,000 times weaker than the electromagnetic.

The weak interaction affects all the fermions of the Standard Model, as well as the Higgs boson; neutrinos interact through gravity and the weak interaction only, and neutrinos were the original reason for the name *weak force*. The weak interaction does not produce bound states (nor does it involve binding energy) – something that gravity does on an astronomical scale, that the electromagnetic force does at the atomic level, and that the strong nuclear force does inside nuclei.

Its most noticeable effect is due to its first unique feature: flavor changing. A neutron, for example, is heavier than a proton (its sister nucleon), but it cannot decay into a proton without changing the flavor (type) of one of its two *down* quarks to an *up* quark. Neither the strong interaction nor electromagnetism permit flavour changing, so this proceeds by weak decay; without weak decay, quark properties such as strangeness and charm (associated with the quarks of the same name) would also be conserved across all interactions. All mesons are unstable because of weak decay. In the process known as beta decay, a *down* quark in the neutron can change into an *up* quark by emitting a virtual W− boson which is then converted into an electron and an electron antineutrino. Another example is the electron capture, a common variant of radioactive decay, wherein a proton and an electron within an atom interact, and are changed to a neutron (an up quark is changed to a down quark) and an electron neutrino is emitted.

Due to the large mass of a boson, weak decay occurs more slowly. Hence, weak decay is much less likely to occur before either strong or electromagnetic decay, as they proceed more rapidly. For example, a neutral pion (which decays electromagnetically) has a life of about 10^{-16} seconds, while a charged pion (which decays through the weak interaction) lives about 10^{-8} seconds, a hundred million times longer. In contrast, a free neutron (which also decays through the weak interaction) lives about 15 minutes.

Weak Isospin and Weak Hypercharge

Left-handed fermions in the Standard Model								
Generation 1			Generation 2			Generation 3		
Fermion	Symbol	Weak isospin	Fermion	Symbol	Weak isospin	Fermion	Symbol	Weak isospin
Electron neutrino	ν_e	+1/2	Muon neutrino	ν_μ	+1/2	Tau neutrino	ν_τ	1/2
Electron	e^-	−1/2	Muon	μ^-	−1/2	Tau	τ^-	−1/2
Up quark	u	+1/2	Charm quark	c	+1/2	Top quark	t	+1/2
Down quark	d	−1/2	Strange quark	s	−1/2	Bottom quark	b	−1/2
All left-handed antiparticles have weak isospin of 0. Right-handed antiparticles have the opposite weak isospin.								

All particles have a property called weak isospin (T_3), which serves as a quantum number and governs how that particle behaves in the weak interaction. Weak isospin plays the same role in the weak interaction as does electric charge in electromagnetism, and color charge in the strong interaction. All fermions have a weak isospin value of either $+\frac{1}{2}$ or $-\frac{1}{2}$. For example, the up quark has a T_3 of $+\frac{1}{2}$ and the down quark $-\frac{1}{2}$. A quark never decays through the weak interaction into a quark of the same T_3: quarks with a T_3 of $+\frac{1}{2}$ decay into quarks with a T_3 of $-\frac{1}{2}$ and vice versa.

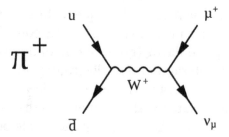

π+ decay through the weak interaction

In any given interaction, weak isospin is conserved: the sum of the weak isospin numbers of the particles entering the interaction equals the sum of the weak isospin numbers of the particles exiting that interaction. For example, a (left-handed) π+, with a weak isospin of 1 normally decays into a νμ (+1/2) and a μ+ (as a right-handed antiparticle, +1/2).

Following the development of the electroweak theory, another property, weak hypercharge, was developed. It is dependent on a particle's electrical charge and weak isospin, and is defined as:

$$Y_W = 2(Q - T_3)$$

where Y_w is the weak hypercharge of a given type of particle, Q is its electrical charge (in elementary charge units) and T_3 is its weak isospin. Whereas some particles have a weak isospin of zero, all particles, except gluons, have non-zero weak hypercharge. Weak hypercharge is the generator of the U(1) component of the electroweak gauge group.

Interaction Types

There are two types of weak interaction (called *vertices*). The first type is called the "charged-current interaction" because it is mediated by particles that carry an electric charge (the W+ or W− bosons), and is responsible for the beta decay phenomenon. The second type is called the "neutral-current interaction" because it is mediated by a neutral particle, the Z boson.

Charged-current Interaction

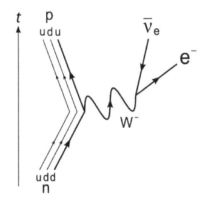

The Feynman diagram for beta-minus decay of a neutron into a proton, electron and electron anti-neutrino, via an intermediate heavy W− boson

In one type of charged current interaction, a charged lepton (such as an electron or a muon, having a charge of −1) can absorb a W+ boson (a particle with a charge of +1) and be thereby converted into a corresponding neutrino (with a charge of 0), where the type ("flavour") of neutrino (electron, muon or tau) is the same as the type of lepton in the interaction, for example:

$$\mu^- + W^+ \rightarrow \nu_\mu$$

Similarly, a down-type quark (d with a charge of $-\frac{1}{3}$) can be converted into an up-type quark (u, with a charge of $+\frac{2}{3}$), by emitting a W− boson or by absorbing a W+ boson. More precisely, the down-type quark becomes a quantum superposition of up-type quarks: that is to say, it has a possibility of becoming any one of the three up-type quarks, with the probabilities given in the CKM matrix tables. Conversely, an up-type quark can emit a W+ boson, or absorb a W− boson, and thereby be converted into a down-type quark, for example:

$$d \rightarrow u + W^-$$

$$d + W^+ \rightarrow u$$

$$c \rightarrow s + W^+$$

$$c + W^- \rightarrow s$$

The W boson is unstable so will rapidly decay, with a very short lifetime. For example:

$$W^- \rightarrow e^- + \bar{v}_e$$

$$W^+ \rightarrow e^+ + v_e$$

Decay of the W boson to other products can happen, with varying probabilities.

In the so-called beta decay of a neutron, a down quark within the neutron emits a virtual W− boson and is thereby converted into an up quark, converting the neutron into a proton. Because of the energy involved in the process (i.e., the mass difference between the down quark and the up quark), the W− boson can only be converted into an electron and an electron-antineutrino. At the quark level, the process can be represented as:

$$d \rightarrow u + e^- + \bar{v}_e$$

Neutral-current Interaction

In neutral current interactions, a quark or a lepton (e.g., an electron or a muon) emits or absorbs a neutral Z boson. For example:

$$e^- \rightarrow e^- + Z^0$$

Like the W boson, the Z boson also decays rapidly, for example:

$$Z^0 \rightarrow b + \bar{b}$$

Electroweak Theory

The Standard Model of particle physics describes the electromagnetic interaction and the weak interaction as two different aspects of a single electroweak interaction, the theory of which was developed around 1968 by Sheldon Glashow, Abdus Salam and Steven Weinberg. They were awarded the 1979 Nobel Prize in Physics for their work. The Higgs mechanism provides an explanation for the presence of three massive gauge bosons (the three carriers of the weak interaction) and the massless photon of the electromagnetic interaction.

According to the electroweak theory, at very high energies, the universe has four mass-less gauge boson fields similar to the photon and a complex scalar Higgs field doublet. However, at low energies, gauge symmetry is spontaneously broken down to the U(1) symmetry of electromagnetism (one of the Higgs fields acquires a vacuum expecta-tion value). This symmetry breaking would produce three massless bosons, but they become integrated by three photon-like fields (through the Higgs mechanism) giving them mass. These three fields become the W+, W− and Z bosons of the weak interac-tion, while the fourth gauge field, which remains massless, is the photon of electromag-netism.

This theory has made a number of predictions, including a prediction of the masses of the Z and W bosons before their discovery. On 4 July 2012, the CMS and the ATLAS experimental teams at the Large Hadron Collider independently announced that they had confirmed the formal discovery of a previously unknown boson of mass between 125–127 GeV/c², whose behaviour so far was "consistent with" a Higgs boson, while adding a cautious note that further data and analysis were needed before positively identifying the new boson as being a Higgs boson of some type. By 14 March 2013, the Higgs boson was tentatively confirmed to exist .

Violation of Symmetry

Left- and right-handed particles: p is the particle's momentum and S is its spin.
Note the lack of reflective symmetry between the states.

The laws of nature were long thought to remain the same under mirror reflection. The results of an experiment viewed via a mirror were expected to be identical to the results of a mirror-reflected copy of the experimental apparatus. This so-called law of parity conservation was known to be respected by classical gravitation, electromagnetism and the strong interaction; it was assumed to be a universal law. However, in the mid-1950s Chen Ning Yang and Tsung-Dao Lee suggested that the weak interaction might violate this law. Chien Shiung Wu and collaborators in 1957 discovered that the weak interac-tion violates parity, earning Yang and Lee the 1957 Nobel Prize in Physics.

Although the weak interaction was once described by Fermi's theory, the discovery of parity violation and renormalization theory suggested that a new approach was needed. In 1957, Robert Marshak and George Sudarshan and, somewhat later, Richard Feyn-man and Murray Gell-Mann proposed a V–A (vector minus axial vector or left-hand-ed) Lagrangian for weak interactions. In this theory, the weak interaction acts only on left-handed particles (and right-handed antiparticles). Since the mirror reflection of a left-handed particle is right-handed, this explains the maximal violation of parity. Interestingly, the V–A theory was developed before the discovery of the Z boson, so it did not include the right-handed fields that enter in the neutral current interaction.

However, this theory allowed a compound symmetry CP to be conserved. CP combines parity P (switching left to right) with charge conjugation C (switching particles with antiparticles). Physicists were again surprised when in 1964, James Cronin and Val Fitch provided clear evidence in kaon decays that CP symmetry could be broken too, winning them the 1980 Nobel Prize in Physics. In 1973, Makoto Kobayashi and Toshihide Maskawa showed that CP violation in the weak interaction required more than two generations of particles, effectively predicting the existence of a then unknown third generation. This discovery earned them half of the 2008 Nobel Prize in Physics. Unlike parity violation, CP violation occurs in only a small number of instances, but remains widely held as an answer to the difference between the amount of matter and antimatter in the universe; it thus forms one of Andrei Sakharov's three conditions for baryogenesis.

Strong Interaction

In particle physics, the strong interaction is the mechanism responsible for the strong nuclear force (also called the strong force, nuclear strong force), one of the four known fundamental interactions, the others are electromagnetism, the weak interaction and gravitation. At the range of a femtometer, it is the strongest force, is approximately 137 times stronger than electromagnetism, a million times stronger than weak interaction and 10^{38} times stronger than gravitation. The strong nuclear force ensures the stability of ordinary matter, confining quarks into hadron particles, such as the proton and neutron, and the further binding of neutrons and protons into atomic nuclei. Most of the mass-energy of a common proton or neutron is in the form of the strong force field energy; the individual quarks provide only about 1% of the mass-energy of a proton.

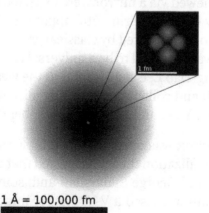

1 Å = 100,000 fm

The nucleus of a helium atom. The two protons have the same charge, but still stay together due to the residual nuclear force

The strong interaction is observable at two ranges: on a larger scale (about 1 to 3 femtometers (fm)), it is the force that binds protons and neutrons (nucleons) together to form the nucleus of an atom. On the smaller scale (less than about 0.8 fm, the radius of a nucleon), it is the force (carried by gluons) that holds quarks together to form protons, neutrons, and other hadron particles. In the latter context, it is often known as the color force. The strong force inherently has such a high strength that hadrons bound by the strong force can produce new massive particles. Thus, if hadrons are struck by high-energy particles, they give rise to new hadrons instead of emitting freely moving radiation (gluons). This property of the strong force is called color confinement, and it prevents the free "emission" of the strong force: instead, in practice, jets of massive particles are observed.

In the context of binding protons and neutrons together to form atomic nuclei, the strong interaction is called the nuclear force (or *residual strong force*). In this case, it is the residuum of the strong interaction between the quarks that make up the protons and neutrons. As such, the residual strong interaction obeys a quite different distance-dependent behavior between nucleons, from when it is acting to bind quarks within nucleons. The binding energy that is partly released on the breakup of a nucleus is related to the residual strong force and is harnessed in nuclear power and fission-type nuclear weapons.

The strong interaction is hypothesized to be mediated by massless particles called gluons, that are exchanged between quarks, antiquarks, and other gluons. Gluons, in turn, are thought to interact with quarks and gluons as all carry a type of charge called color charge. Color charge is analogous to electromagnetic charge, but it comes in three types rather than one (+/- red, +/- green, +/- blue) that results in a different type of force, with different rules of behavior. These rules are detailed in the theory of quantum chromodynamics (QCD), which is the theory of quark-gluon interactions.

After the Big Bang, during the electroweak epoch, the electroweak force separated from the strong force. A Grand Unified Theory is hypothesized to exist to describe this, no such theory has been successfully formulated yet, and the unification remains an unsolved problem in physics.

History

Before the 1970s, physicists were uncertain as to how the atomic nucleus was bound together. It was known that the nucleus was composed of protons and neutrons and that protons possessed positive electric charge, while neutrons were electrically neutral. By the understanding of physics at that time, positive charges would repel one another and the positively charged protons should cause the nucleus to fly apart. However, this was never observed. New physics was needed to explain this phenomenon.

A stronger attractive force was postulated to explain how the atomic nucleus was bound despite the protons' mutual electromagnetic repulsion. This hypothesized force was called the *strong force*, which was believed to be a fundamental force that acted on the protons and neutrons that make up the nucleus.

It was later discovered that protons and neutrons were not fundamental particles, but were made up of constituent particles called quarks. The strong attraction between nucleons was the side-effect of a more fundamental force that bound the quarks together into protons and neutrons. The theory of quantum chromodynamics explains that quarks carry what is called a color charge, although it has no relation to visible color. Quarks with unlike color charge attract one another as a result of the strong interaction, and the particle that mediated this was called the gluon.

Details

The fundamental couplings of the strong interaction, from left to right: gluon radiation, gluon splitting and gluon self-coupling.

The word *strong* is used since the strong interaction is the "strongest" of the four fundamental forces; its strength is around 137 times that of the electromagnetic force, some 10^6 times as great as that of the weak force, and about 10^{38} times that of gravitation, at a distance of 10^{-15} meter (femtometer) or less.

Behaviour of the Strong Force

The strong force is described by quantum chromodynamics (QCD), a part of the standard model of particle physics. Mathematically, QCD is a non-Abelian gauge theory based on a local (gauge) symmetry group called SU(3).

Quarks and gluons are the only fundamental particles that carry non-vanishing color charge, and hence participate in strong interactions. The strong force itself acts directly only on elementary quark and gluon particles.

All quarks and gluons in QCD interact with each other through the strong force. The strength of interaction is parametrized by the strong coupling constant. This strength is modified by the gauge color charge of the particle, a group theoretical property.

The strong force acts between quarks. Unlike all other forces (electromagnetic, weak, and gravitational), the strong force does not diminish in strength with increasing distance. After a limiting distance (about the size of a hadron) has been reached, it remains at a strength of about 10,000 newtons, no matter how much farther the distance between the quarks. In QCD, this phenomenon is called color confinement; as a result only hadrons, not individual free quarks, can be observed. The explanation is that the amount of work done against a force of 10,000 newtons is enough to create particle-antiparticle pairs within a very short distance of that interaction. In simple terms, the very energy added to the system required to pull two quarks apart would create a pair of new quarks that will pair up with the original ones. The failure of all experiments that have searched for free quarks is considered to be evidence of this phenomenon.

The elementary quark and gluon particles involved in a high energy collision are not directly observable. They instead emerge as jets of newly created hadrons, whenever sufficient energy is deposited into a quark-quark bond, as when a quark in one proton is struck by a very fast quark of another impacting proton during a particle accelerator experiment. However, quark–gluon plasmas have been observed.

Every quark in the universe does not attract every other quark in the above distance independent manner, since color-confinement implies that the strong force acts without distance-diminishment only between pairs of single quarks, and that in collections of bound quarks (i.e., hadrons), the net color-charge of the quarks cancels out, resulting in a limit of the action of the forces. Collections of quarks (hadrons) therefore appear nearly without color-charge, and the strong force is therefore nearly absent between those hadrons (i.e., between baryons or mesons). However, the cancellation is not quite perfect. A small residual force remains (described below) known as the residual strong force. This residual force *does* diminish rapidly with distance, and is thus very short-range (effectively a few femtometers). It manifests as a force between the "colorless" hadrons, and is sometimes known as the strong nuclear force or simply nuclear force.

Residual Strong Force

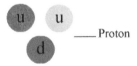

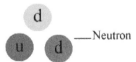

An animation of the nuclear force (or residual strong force) interaction between a proton and a neutron. The small colored double circles are gluons, which can be seen binding the proton and neutron together. These gluons also hold the quark-antiquark combination called the pion together, and thus help transmit a residual part of the strong force even between colorless hadrons. Anticolors are shown as per this diagram. For a larger version, click here

The residual effect of the strong force is called the nuclear force. The nuclear force acts between hadrons, such as mesons or the baryons (such as nucleons) in atomic nuclei. This "residual strong force", acting indirectly, transmits gluons that form part of the virtual pi and rho mesons, which, in turn, transmit the nuclear force between nucleons.

The residual strong force is thus a minor residuum of the strong force that binds quarks together into protons and neutrons. This same force is much weaker *between* neutrons and protons, because it is mostly neutralized *within* them, in the same way that electromagnetic forces between neutral atoms (van der Waals forces) are much weaker than the electromagnetic forces that hold the atoms internally together.

Unlike the strong force itself, the nuclear force, or residual strong force, *does* diminish in strength, and it in fact diminishes rapidly with distance. The decrease is approximately as a negative exponential power of distance, though there is no simple expression known for this; see Yukawa potential. The rapid decrease with distance of the attractive residual force and the less-rapid decrease of the repulsive electromagnetic force acting between protons within a nucleus, causes the instability of larger atomic nuclei, such as all those with atomic numbers larger than 82 (the element lead).

Electroweak Interaction

In particle physics, the electroweak interaction is the unified description of two of the four known fundamental interactions of nature: electromagnetism and the weak interaction. Although these two forces appear very different at everyday low energies, the theory models them as two different aspects of the same force. Above the unification energy, on the order of 100 GeV, they would merge into a single electroweak force. Thus, if the universe is hot enough (approximately 10^{15} K, a temperature exceeded until shortly after the Big Bang), then the electromagnetic force and weak force merge into a combined electroweak force. During the electroweak epoch, the electroweak force separated from the strong force. During the quark epoch, the electroweak force split into the electromagnetic and weak force.

Sheldon Glashow, Abdus Salam, and Steven Weinberg were awarded the 1979 Nobel Prize in Physics for their contributions to the unification of the weak and electromagnetic interaction between elementary particles. The existence of the electroweak interactions was experimentally established in two stages, the first being the discovery of neutral currents in neutrino scattering by the Gargamelle collaboration in 1973, and the second in 1983 by the UA1 and the UA2 collaborations that involved the discovery of the W and Z gauge bosons in proton–antiproton collisions at the converted Super Proton Synchrotron. In 1999, Gerardus 't Hooft and Martinus Veltman were awarded the Nobel prize for showing that the electroweak theory is renormalizable.

Formulation

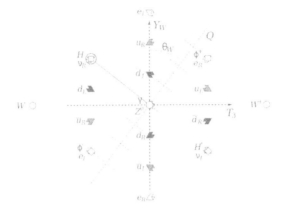

The pattern of weak isospin, T_3, and weak hypercharge, Y_W, of the known elementary particles, showing electric charge, Q, along the weak mixing angle. The neutral Higgs field (circled) breaks the electroweak symmetry and interacts with other particles to give them mass. Three components of the Higgs field become part of the massive W and Z bosons.

Mathematically, the unification is accomplished under an $SU(2) \times U(1)$ gauge group. The corresponding gauge bosons are the three W bosons of weak isospin from SU(2) (W_1, W_2, and W_3), and the B boson of weak hypercharge from U(1), respectively, all of which are massless.

In the Standard Model, the W± and Zo bosons, and the photon, are produced by the spontaneous symmetry breaking of the electroweak symmetry from $SU(2) \times U(1)_Y$ to $U(1)_{em}$, caused by the Higgs mechanism. $U(1)_Y$ and $U(1)_{em}$ are different copies of $U(1)$; the generator of $U(1)_{em}$ is given by $Q = Y/2 + I_3$, where Y is the generator of $U(1)_Y$ (called the weak hypercharge), and I_3 is one of the $SU(2)$ generators (a component of weak isospin).

The spontaneous symmetry breaking causes the W_3 and B bosons to coalesce together into two different bosons – the Zo boson, and the photon (γ) as follows:

$$\begin{pmatrix} \gamma \\ Z^0 \end{pmatrix} = \begin{pmatrix} \cos\theta_W & \sin\theta_W \\ -\sin\theta_W & \cos\theta_W \end{pmatrix} \begin{pmatrix} B \\ W_3 \end{pmatrix}$$

Where θ_W is the *weak mixing angle*. The axes representing the particles have essentially just been rotated, in the (W_3, B) plane, by the angle θ_W. This also introduces a discrepancy between the mass of the Zo and the mass of the W± particles (denoted as M_Z and M_W, respectively);

$$M_Z = \frac{M_W}{\cos\theta_W}$$

W_1 and W_2 bosons, in turn, combine to give massive charged bosons

$$W^{\pm} = \frac{1}{\sqrt{2}}(W_1 \mp iW_2)$$

The distinction between electromagnetism and the weak force arises because there is a (nontrivial) linear combination of Y and I_3 that vanishes for the Higgs boson (it is an eigenstate of both Y and I_3, so the coefficients may be taken as $-I_3$ and Y): $U(1)_{em}$ is defined to be the group generated by this linear combination, and is unbroken because it does not interact with the Higgs.

Lagrangian

Before Electroweak Symmetry Breaking

The Lagrangian for the electroweak interactions is divided into four parts before electroweak symmetry breaking

$$\mathcal{L}_{EW} = \mathcal{L}_g + \mathcal{L}_f + \mathcal{L}_h + \mathcal{L}_y$$

The $\mathcal{L}_g$ term describes the interaction between the three W particles and the B particle.

$$\mathcal{L}_g = -\frac{1}{4}W_a^{\mu\nu}W_{\mu\nu}^a - \frac{1}{4}B^{\mu\nu}B_{\mu\nu},$$

where $W^{a\mu\nu} (a = 1, 2, 3)$ and $B^{\mu\nu}$ are the field strength tensors for the weak isospin and weak hypercharge fields.

$\mathcal{L}_f$ is the kinetic term for the Standard Model fermions. The interaction of the gauge bosons and the fermions are through the gauge covariant derivative.

$$\mathcal{L}_f = \bar{Q}_i i D\!\!\!/\, Q_i + \bar{u}_i i D\!\!\!/\, u_i + \bar{d}_i i D\!\!\!/\, d_i + \bar{L}_i i D\!\!\!/\, L_i + \bar{e}_i i D\!\!\!/\, e_i,$$

where the subscript runs over the three generations of fermions, Q, u, and d are the left-handed doublet, right-handed singlet up, and right handed singlet down quark fields, and L and e are the left-handed doublet and right-handed singlet electron fields.

The h term describes the Higgs field F.

$$\mathcal{L}_h = |D_\mu h|^2 - \lambda\left(|h|^2 - \frac{v^2}{2}\right)^2$$

The y term gives the Yukawa interaction that generates the fermion masses after the Higgs acquires a vacuum expectation value.

$$\mathcal{L}_y = -y_{uij}\epsilon^{ab} h_b^\dagger \bar{Q}_{ia} u_j^c - y_{dij} h \bar{Q}_i d_j^c - y_{eij} h \bar{L}_i e_j^c + h.c.$$

After Electroweak Symmetry Breaking

The Lagrangian reorganizes itself after the Higgs boson acquires a vacuum expectation value. Due to its complexity, this Lagrangian is best described by breaking it up into several parts as follows.

$$\mathcal{L}_{EW} = \mathcal{L}_K + \mathcal{L}_N + \mathcal{L}_C + \mathcal{L}_H + \mathcal{L}_{HV} + \mathcal{L}_{WWV} + \mathcal{L}_{WWVV} + \mathcal{L}_Y$$

The kinetic term $\mathcal{L}_K$ contains all the quadratic terms of the Lagrangian, which include the dynamic terms (the partial derivatives) and the mass terms (conspicuously absent from the Lagrangian before symmetry breaking)

$$\mathcal{L}_K = \sum_f \overline{f}(i\partial\!\!\!/ - m_f)f - \frac{1}{4}A_{\mu\nu}A^{\mu\nu} - \frac{1}{2}W^+_{\mu\nu}W^{-\mu\nu} + m^2_W W^+_\mu W^{-\mu}$$

$$- \frac{1}{4}Z_{\mu\nu}Z^{\mu\nu} + \frac{1}{2}m^2_Z Z_\mu Z^\mu + \frac{1}{2}(\partial^\mu H)(\partial_\mu H) - \frac{1}{2}m^2_H H^2$$

where the sum runs over all the fermions of the theory (quarks and leptons), and the fields $A_{\mu\nu}$, $Z_{\mu\nu}$, $W^-_{\mu\nu}$, and $W^+_{\mu\nu} \equiv (W^-_{\mu\nu})^\dagger$ are given as

$X_{\mu\nu} = \partial_\mu X_\nu - \partial_\nu X_\mu + gf^{abc}X^b_\mu X^c_\nu$, (replace X by the relevant field, and f^{abc} with the structure constants for the gauge group).

The neutral current $\mathcal{L}_N$ and charged current $\mathcal{L}_C$ components of the Lagrangian contain the interactions between the fermions and gauge bosons.

$$\mathcal{L}_N = eJ^{em}_\mu A^\mu + \frac{g}{\cos\theta_W}(J^3_\mu - \sin^2\theta_W J^{em}_\mu)Z^\mu,$$

where the electromagnetic current J^{em}_μ and the neutral weak current J^3_μ are

$$J^{em}_\mu = \sum_f q_f \overline{f}\gamma_\mu f,$$

and

$$J^3_\mu = \sum_f I^3_f \overline{f}\gamma_\mu \frac{1-\gamma^5}{2}f$$

q_f and I^3_f are the fermions' electric charges and weak isospin.

The charged current part of the Lagrangian is given by

$$\mathcal{L}_C = -\frac{g}{\sqrt{2}}\left[\overline{u}_i\gamma^\mu \frac{1-\gamma^5}{2}M^{CKM}_{ij}d_j + \overline{v}_i\gamma^\mu \frac{1-\gamma^5}{2}e_i\right]W^+_\mu + h.c.$$

$\mathcal{L}_H$ contains the Higgs three-point and four-point self interaction terms.

$$\mathcal{L}_H = -\frac{gm_H^2}{4m_W}H^3 - \frac{g^2 m_H^2}{32 m_W^2}H^4$$

$\mathcal{L}_{HV}$ contains the Higgs interactions with gauge vector bosons.

$$\mathcal{L}_{HV} = \left(gm_W H + \frac{g^2}{4}H^2 \right)\left(W_\mu^+ W^{-\mu} + \frac{1}{2\cos^2\theta_W}Z_\mu Z^\mu \right)$$

$\mathcal{L}_{WWV}$ contains the gauge three-point self interactions.

$$\mathcal{L}_{WWV} = -ig[(W_{\mu\nu}^+ W^{-\mu} - W^{+\mu}W_{\mu\nu}^-)(A^\nu \sin\theta_W - Z^\nu \cos\theta_W) + W_\nu^- W_\mu^+ (A^{\mu\nu}\sin\theta_W - Z^{\mu\nu}\cos\theta_W)]$$

$\mathcal{L}_{WWVV}$ contains the gauge four-point self interactions

$$\mathcal{L}_{WWVV} = -\frac{g^2}{4}\Big\{ \ [2W_\mu^+ W^{-\mu} + (A_\mu \sin\theta_W - Z_\mu \cos\theta_W)^2]^2$$

$$-[W_\mu^+ W_\nu^- + W_\nu^+ W_\mu^- + (A_\mu \sin\theta_W - Z_\mu \cos\theta_W)(A_\nu \sin\theta_W - Z_\nu \cos\theta_W)]^2 \Big\}$$

and $\mathcal{L}_Y$ contains the Yukawa interactions between the fermions and the Higgs field.

$$\mathcal{L}_Y = -\sum_f \frac{gm_f}{2m_W}\bar{f}fH$$

Note the $\dfrac{1-\gamma^5}{2}$ factors in the weak couplings: these factors project out the left handed components of the spinor fields. This is why electroweak theory (after symmetry breaking) is commonly said to be a chiral theory.

References

* John Gribbin (1998). Q is for Quantum: Particle Physics from A to Z. London: Weidenfeld & Nicolson. p. 138. ISBN 0-297-81752-3.

* Richard Feynman (1970). The Feynman Lectures on Physics Vol II. Addison Wesley Longman. ISBN 978-0-201-02115-8.

* M. Mansfield; C. O'Sullivan (2011). Understanding Physics (4th ed.). John Wiley & Sons. ISBN 978-0-47-0746370.

* Peskin, Michael E.; Schroeder, Daniel V. (1995). An Introduction to Quantum Fields. Westview Press. p. 198. ISBN 0-201-50397-2.

* R. Resnick; R. Eisberg (1985). Quantum Physics of Atoms, Molecules, Solids, Nuclei and Particles (2nd ed.). John Wiley & Sons. p. 684. ISBN 978-0-471-87373-0.

* Giachetta, G., Mangiarotti, L., Sardanashvily, G. (2009) Advanced Classical Field Theory. Singapore: World Scientific, ISBN 978-981-283-895-7.

- Peter Watkins (1986). Story of the W and Z. Cambridge: Cambridge University Press. p. 70. ISBN 978-0-521-31875-4.

- Paul Langacker (2001) [1989]. "Cp Violation and Cosmology". In Cecilia Jarlskog. CP violation. London, River Edge: World Scientific Publishing Co. p. 552. ISBN 9789971505615.

- Feynman, R. P. (1985). QED: The Strange Theory of Light and Matter. Princeton University Press. p. 136. ISBN 0-691-08388-6.

- Walter Greiner; Berndt Müller (2009). Gauge Theory of Weak Interactions. Springer. p. 2. ISBN 978-3-540-87842-1.

- Charles W. Carey (2006). "Lee, Tsung-Dao". American scientists. Facts on File Inc. p. 225. ISBN 9781438108070.

3

Theories of Particle Physics

The theories of particle physics discussed in this chapter are quantum chromodynamics, quantum field theory, gauge theory and peccei-quinn theory. Quantum chromodynamics is the theory of strong interactions whereas quantum field theory framework is a part of particle physics that is used for constructing quantum mechanical models of subatomic particles. The aim of this section is to explore the theories of particle physics.

Quantum Chromodynamics

In theoretical physics, quantum chromodynamics (QCD) is the theory of strong interactions, a fundamental force describing the interactions between quarks and gluons which make up hadrons such as the proton, neutron and pion. QCD is a type of quantum field theory called a non-abelian gauge theory with symmetry group SU(3). The QCD analog of electric charge is a property called *color*. Gluons are the force carrier of the theory, like photons are for the electromagnetic force in quantum electrodynamics. The theory is an important part of the Standard Model of particle physics. A large body of experimental evidence for QCD has been gathered over the years.

QCD enjoys two peculiar properties:

- Confinement, which means that the force between quarks does not diminish as they are separated. Because of this, when you do separate a quark from other quarks, the energy in the gluon field is enough to create another quark pair; they are thus forever bound into hadrons such as the proton and the neutron or the pion and kaon. Although analytically unproven, confinement is widely believed to be true because it explains the consistent failure of free quark searches, and it is easy to demonstrate in lattice QCD.

- Asymptotic freedom, which means that in very high-energy reactions, quarks and gluons interact very weakly creating a quark–gluon plasma. This prediction of QCD was first discovered in the early 1970s by David Politzer, Frank Wilczek and David Gross. For this work they were awarded the 2004 Nobel Prize in Physics.

The phase transition temperature between these two properties has been measured by the ALICE experiment to be well above 160 MeV. Below this temperature, confinement is dominant, while above it, asymptotic freedom becomes dominant.

Terminology

The word *quark* was coined by American physicist Murray Gell-Mann (b. 1929) in its present sense. It originally comes from the phrase "Three quarks for Muster Mark" in *Finnegans Wake* by James Joyce. On June 27, 1978, Gell-Mann wrote a private letter to the editor of the *Oxford English Dictionary*, in which he related that he had been influenced by Joyce's words: "The allusion to three quarks seemed perfect." (Originally, only three quarks had been discovered.) Gell-Mann, however, wanted to pronounce the word to rhyme with "fork" rather than with "park", as Joyce seemed to indicate by rhyming words in the vicinity such as *Mark*. Gell-Mann got around that "by supposing that one ingredient of the line 'Three quarks for Muster Mark' was a cry of 'Three quarts for Mister ...' heard in H.C. Earwicker's pub", a plausible suggestion given the complex punning in Joyce's novel.

The three kinds of charge in QCD (as opposed to one in quantum electrodynamics or QED) are usually referred to as "color charge" by loose analogy to the three kinds of color (red, green and blue) perceived by humans. Other than this nomenclature, the quantum parameter "color" is completely unrelated to the everyday, familiar phenomenon of color.

Since the theory of electric charge is dubbed "electrodynamics", the Greek word "chroma" (meaning color) is applied to the theory of color charge, "chromodynamics".

History

With the invention of bubble chambers and spark chambers in the 1950s, experimental particle physics discovered a large and ever-growing number of particles called hadrons. It seemed that such a large number of particles could not all be fundamental. First, the particles were classified by charge and isospin by Eugene Wigner and Werner Heisenberg; then, in 1953, according to strangeness by Murray Gell-Mann and Kazuhiko Nishijima. To gain greater insight, the hadrons were sorted into groups having similar properties and masses using the *eightfold way*, invented in 1961 by Gell-Mann and Yuval Ne'eman. Gell-Mann and George Zweig, correcting an earlier approach of Shoichi Sakata, went on to propose in 1963 that the structure of the groups could be explained by the existence of three flavors of smaller particles inside the hadrons: the quarks.

Perhaps the first remark that quarks should possess an additional quantum number was made as a short footnote in the preprint of Boris Struminsky in connection with Ω^- hyperon composed of three strange quarks with parallel spins (this situation was peculiar, because since quarks are fermions, such combination is forbidden by the Pauli exclusion principle):

Three identical quarks cannot form an antisymmetric S-state. In order to realize an antisymmetric orbital S-state, it is necessary for the quark to have an additional quantum number.

—*B. V. Struminsky, Magnetic moments of barions in the quark model, JINR-Preprint P-1939, Dubna, Submitted on January 7, 1965*

Boris Struminsky was a PhD student of Nikolay Bogolyubov. The problem considered in this preprint was suggested by Nikolay Bogolyubov, who advised Boris Struminsky in this research. In the beginning of 1965, Nikolay Bogolyubov, Boris Struminsky and Albert Tavkhelidze wrote a preprint with a more detailed discussion of the additional quark quantum degree of freedom. This work was also presented by Albert Tavchelidze without obtaining consent of his collaborators for doing so at an international conference in Trieste (Italy), in May 1965.

A similar mysterious situation was with the Δ^{++} baryon; in the quark model, it is composed of three up quarks with parallel spins. In 1965, Moo-Young Han with Yoichiro Nambu and Oscar W. Greenberg independently resolved the problem by proposing that quarks possess an additional SU(3) gauge degree of freedom, later called color charge. Han and Nambu noted that quarks might interact via an octet of vector gauge bosons: the gluons.

Since free quark searches consistently failed to turn up any evidence for the new particles, and because an elementary particle back then was *defined* as a particle which could be separated and isolated, Gell-Mann often said that quarks were merely convenient mathematical constructs, not real particles. The meaning of this statement was usually clear in context: He meant quarks are confined, but he also was implying that the strong interactions could probably not be fully described by quantum field theory.

Richard Feynman argued that high energy experiments showed quarks are real particles: he called them *partons* (since they were parts of hadrons). By particles, Feynman meant objects which travel along paths, elementary particles in a field theory.

The difference between Feynman's and Gell-Mann's approaches reflected a deep split in the theoretical physics community. Feynman thought the quarks have a distribution of position or momentum, like any other particle, and he (correctly) believed that the diffusion of parton momentum explained diffractive scattering. Although Gell-Mann believed that certain quark charges could be localized, he was open to the possibility that the quarks themselves could not be localized because space and time break down. This was the more radical approach of S-matrix theory.

James Bjorken proposed that pointlike partons would imply certain relations should hold in deep inelastic scattering of electrons and protons, which were spectacularly verified in experiments at SLAC in 1969. This led physicists to abandon the S-matrix approach for the strong interactions.

The discovery of asymptotic freedom in the strong interactions by David Gross, David Politzer and Frank Wilczek allowed physicists to make precise predictions of the results of many high energy experiments using the quantum field theory technique of perturbation theory. Evidence of gluons was discovered in three-jet events at PETRA in 1979. These experiments became more and more precise, culminating in the verification of perturbative QCD at the level of a few percent at the LEP in CERN.

The other side of asymptotic freedom is confinement. Since the force between color charges does not decrease with distance, it is believed that quarks and gluons can never be liberated from hadrons. This aspect of the theory is verified within lattice QCD computations, but is not mathematically proven. One of the Millennium Prize Problems announced by the Clay Mathematics Institute requires a claimant to produce such a proof. Other aspects of non-perturbative QCD are the exploration of phases of quark matter, including the quark–gluon plasma.

The relation between the short-distance particle limit and the confining long-distance limit is one of the topics recently explored using string theory, the modern form of S-matrix theory.

Theory

?	**Unsolved problem in physics**: *QCD in the non-perturbative regime:* • Confinement: the equations of QCD remain unsolved at energy scales relevant for describing atomic nuclei. How does QCD give rise to the physics of nuclei and nuclear constituents? • Quark matter: the equations of QCD predict that a plasma (or soup) of quarks and gluons should be formed at high temperature and density. What are the properties of this phase of matter? *(more unsolved problems in physics)*

Some Definitions

Every field theory of particle physics is based on certain symmetries of nature whose existence is deduced from observations. These can be

- local symmetries, that are the symmetries that act independently at each point in spacetime. Each such symmetry is the basis of a gauge theory and requires the introduction of its own gauge bosons.

- global symmetries, which are symmetries whose operations must be simultaneously applied to all points of spacetime.

QCD is a gauge theory of the SU(3) gauge group obtained by taking the color charge to define a local symmetry.

Since the strong interaction does not discriminate between different flavors of quark, QCD has approximate flavor symmetry, which is broken by the differing masses of the quarks.

There are additional global symmetries whose definitions require the notion of chirality, discrimination between left and right-handed. If the spin of a particle has a positive projection on its direction of motion then it is called left-handed; otherwise, it is right-handed. Chirality and handedness are not the same, but become approximately equivalent at high energies.

- Chiral symmetries involve independent transformations of these two types of particle.

- Vector symmetries (also called diagonal symmetries) mean the same transformation is applied on the two chiralities.

- Axial symmetries are those in which one transformation is applied on left-handed particles and the inverse on the right-handed particles.

Additional Remarks: Duality

As mentioned, *asymptotic freedom* means that at large energy – this corresponds also to *short distances* – there is practically no interaction between the particles. This is in contrast – more precisely one would say *dual* – to what one is used to, since usually one connects the absence of interactions with *large* distances. However, as already mentioned in the original paper of Franz Wegner, a solid state theorist who introduced 1971 simple gauge invariant lattice models, the high-temperature behaviour of the *original model*, e.g. the strong decay of correlations at large distances, corresponds to the low-temperature behaviour of the (usually ordered!) *dual model*, namely the asymptotic decay of non-trivial correlations, e.g. short-range deviations from almost perfect arrangements, for short distances. Here, in contrast to Wegner, we have only the dual model, which is that one described in this article.

Symmetry Groups

The color group SU(3) corresponds to the local symmetry whose gauging gives rise to QCD. The electric charge labels a representation of the local symmetry group U(1) which is gauged to give QED: this is an abelian group. If one considers a version of QCD with N_f flavors of massless quarks, then there is a global (chiral) flavor symmetry group $SU_L(N_f) \times SU_R(N_f) \times U_B(1) \times U_A(1)$. The chiral symmetry is spontaneously broken by the QCD vacuum to the vector (L+R) $SU_V(N_f)$ with the formation of a chiral condensate. The vector symmetry, $U_B(1)$ corresponds to the baryon number of quarks and is an exact symmetry. The axial symmetry $U_A(1)$ is exact in the classical theory, but broken in the quantum theory, an occurrence called an anomaly. Gluon field configurations called instantons are closely related to this anomaly.

There are two different types of SU(3) symmetry: there is the symmetry that acts on the different colors of quarks, and this is an exact gauge symmetry mediated by the gluons, and there is also a flavor symmetry which rotates different flavors of quarks to each other, or *flavor SU(3)*. Flavor SU(3) is an approximate symmetry of the vacuum of QCD, and is not a fundamental symmetry at all. It is an accidental consequence of the small mass of the three lightest quarks.

In the QCD vacuum there are vacuum condensates of all the quarks whose mass is less than the QCD scale. This includes the up and down quarks, and to a lesser extent the strange quark, but not any of the others. The vacuum is symmetric under SU(2) isospin rotations of up and down, and to a lesser extent under rotations of up, down and strange, or full flavor group SU(3), and the observed particles make isospin and SU(3) multiplets.

The approximate flavor symmetries do have associated gauge bosons, observed particles like the rho and the omega, but these particles are nothing like the gluons and they are not massless. They are emergent gauge bosons in an approximate string description of QCD.

Lagrangian

The dynamics of the quarks and gluons are controlled by the quantum chromodynamics Lagrangian. The gauge invariant QCD Lagrangian is

$$\mathcal{L}_{QCD} = \bar{\psi}_i \left(i(\gamma^\mu D_\mu)_{ij} - m\delta_{ij} \right) \psi_j - \frac{1}{4} G^a_{\mu\nu} G^{\mu\nu}_a$$

where $\psi_i(x)$ is the quark field, a dynamical function of spacetime, in the fundamental representation of the SU(3) gauge group, indexed by $i, j, \ldots$; $A^a_\mu(x)$ are the gluon fields, also dynamical functions of spacetime, in the adjoint representation of the SU(3) gauge group, indexed by $a, b, \ldots$ The γ^μ are Dirac matrices connecting the spinor representation to the vector representation of the Lorentz group.

The symbol $G^a_{\mu\nu}$ represents the gauge invariant gluon field strength tensor, analogous to the electromagnetic field strength tensor, $F^{\mu\nu}$, in quantum electrodynamics. It is given by:

$$G^a_{\mu\nu} = \partial_\mu A^a_\nu - \partial_\nu A^a_\mu + g f^{abc} A^b_\mu A^c_\nu$$

where f_{abc} are the structure constants of SU(3). Note that the rules to move-up or pull-down the a, b, or c indices are *trivial*, $(+, \ldots, +)$, so that $f^{abc} = f_{abc} = f^a_{\ bc}$ whereas for the μ or ν indices one has the non-trivial *relativistic* rules, corresponding e.g. to the metric signature $(+ - - -)$.

The constants m and g control the quark mass and coupling constants of the theory, subject to renormalization in the full quantum theory.

An important theoretical notion concerning the final term of the above Lagrangian is the *Wilson loop* variable. This loop variable plays an important role in discretized forms of the QCD, and more generally, it distinguishes confined and deconfined states of a gauge theory. It was introduced by the Nobel prize winner Ken-neth G. Wilson and is treated in a separate article.

Fields

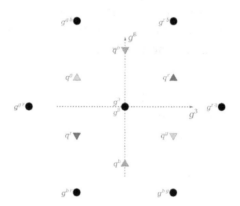

The pattern of strong charges for the three colors of quark, three antiquarks, and eight gluons (with two of zero charge overlapping).

Quarks are massive spin-1/2 fermions which carry a color charge whose gauging is the content of QCD. Quarks are represented by Dirac fields in the fundamental representation 3 of the gauge group SU(3). They also carry electric charge (either −1/3 or 2/3) and participate in weak interactions as part of weak isospin doublets. They carry global quantum numbers including the baryon number, which is 1/3 for each quark, hypercharge and one of the flavor quantum numbers.

Gluons are spin-1 bosons which also carry color charges, since they lie in the adjoint representation 8 of SU(3). They have no electric charge, do not participate in the weak interactions, and have no flavor. They lie in the singlet representation 1 of all these symmetry groups.

Every quark has its own antiquark. The charge of each antiquark is exactly the opposite of the corresponding quark.

Dynamics

According to the rules of quantum field theory, and the associated Feynman diagrams, the above theory gives rise to three basic interactions: a quark may emit (or absorb) a gluon, a gluon may emit (or absorb) a gluon, and two gluons may directly interact. This contrasts with QED, in which only the first kind of interaction occurs, since photons have no charge. Diagrams involving Faddeev–Popov ghosts must be considered too (except in the unitarity gauge).

Area Law and Confinement

Detailed computations with the above-mentioned Lagrangian show that the effective potential between a quark and its anti-quark in a meson contains a term $\propto r$, which represents some kind of "stiffness" of the interaction between the particle and its anti-particle at large distances, similar to the entropic elasticity of a rubber band. This leads to *confinement* of the quarks to the interior of hadrons, i.e. me-sons and nucleons, with typical radii R_c, corresponding to former "Bag models" of the hadrons . The order of magnitude of the "bag radius" is 1 fm (= 10^{-15} m). Moreover, the above-mentioned stiffness is quantitatively related to the so-called "area law" behaviour of the expectation value of the Wilson loop product P of the ordered coupling constants around a closed loop W; i.e. $\langle P_W \rangle$ is proportional to the *area* enclosed by the loop. For this behaviour the non-abelian behaviour of the gauge group is essential.

Methods

Further analysis of the content of the theory is complicated. Various techniques have been developed to work with QCD. Some of them are discussed briefly below.

Perturbative QCD

This approach is based on asymptotic freedom, which allows perturbation theory to be used accurately in experiments performed at very high energies. Although limited in scope, this approach has resulted in the most precise tests of QCD to date.

Lattice QCD

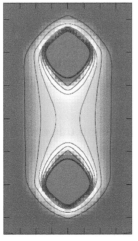

A quark and an antiquark (red color) are glued together (green color) to form a meson
(result of a lattice QCD simulation by M. Cardoso et al.)

Among non-perturbative approaches to QCD, the most well established one is lattice QCD. This approach uses a discrete set of spacetime points (called the lattice) to reduce the analytically intractable path integrals of the continuum theory to a very difficult numerical computation which is then carried out on supercomputers like the QCDOC which was constructed for precisely this purpose. While it is a slow and resource-intensive approach, it has wide applicability, giving insight into parts of the theory inaccessible by other means, in particular into the explicit forces acting between quarks and antiquarks in a meson. However, the numerical sign problem makes it difficult to use lattice methods to study QCD at high density and low temperature (e.g. nuclear matter or the interior of neutron stars).

1/N Expansion

A well-known approximation scheme, the 1/N expansion, starts from the premise that the number of colors is infinite, and makes a series of corrections to account for the fact that it is not. Until now, it has been the source of qualitative insight rather than a method for quantitative predictions. Modern variants include the AdS/CFT approach.

Effective Theories

For specific problems effective theories may be written down which give qualitatively correct results in certain limits. In the best of cases, these may then be obtained as systematic expansions in some parameter of the QCD Lagrangian. One such effective field theory is chiral perturbation theory or ChiPT, which is the QCD effective theory at low energies. More precisely, it is a low energy expansion based on the spontaneous chiral symmetry breaking of QCD, which is an exact symmetry when quark masses are equal to zero, but for the u,d and s quark, which have small mass, it is still a good approximate symmetry. Depending on the number of quarks which are treated as light, one uses either SU(2) ChiPT or SU(3) ChiPT . Other effective theories are heavy quark effective theory (which expands around heavy quark mass near infinity), and soft-collinear effective theory (which expands around large ratios of energy scales). In addition to effective theories, models like the Nambu–Jona-Lasinio model and the chiral model are often used when discussing general features.

QCD Sum Rules

Based on an Operator product expansion one can derive sets of relations that connect different observables with each other.

Nambu–jona-lasinio Model

In one of his recent works, Kei-Ichi Kondo derived as a low-energy limit of QCD, a theory linked to the Nambu–Jona-Lasinio model since it is basically a particular non-local version of the Polyakov–Nambu–Jona-Lasinio model. The later being in its local version, nothing but the Nambu–Jona-Lasinio model in which one has included the Polyakov loop effect, in order to describe a 'certain confinement'.

The Nambu–Jona-Lasinio model in itself is, among many other things, used because it is a 'relatively simple' model of chiral symmetry breaking, phenomenon present up to certain conditions (Chiral limit i.e. massless fermions) in QCD itself. In this model, however, there is no confinement. In particular, the energy of an isolated quark in the physical vacuum turns out well defined and finite.

Experimental Tests

The notion of quark flavors was prompted by the necessity of explaining the properties of hadrons during the development of the quark model. The notion of color was necessitated by the puzzle of the Δ++. This has been dealt with in the section on the history of QCD.

The first evidence for quarks as real constituent elements of hadrons was obtained in deep inelastic scattering experiments at SLAC. The first evidence for gluons came in three jet events at PETRA.

Several good quantitative tests of perturbative QCD exist:

- The running of the QCD coupling as deduced from many observations

- Scaling violation in polarized and unpolarized deep inelastic scattering

- Vector boson production at colliders (this includes the Drell-Yan process)

- Jet cross sections in colliders

- Event shape observables at the LEP

- Heavy-quark production in colliders

Quantitative tests of non-perturbative QCD are fewer, because the predictions are harder to make. The best is probably the running of the QCD coupling as probed through lattice computations of heavy-quarkonium spectra. There is a recent claim about the mass of the heavy meson B_c. Other non-perturbative tests are currently at the level of 5% at best. Continuing work on masses and form factors of hadrons and their weak matrix elements are promising candidates for future quantitative tests. The whole subject of quark matter and the quark–gluon plasma is a non-perturbative test bed for QCD which still remains to be properly exploited.

One qualitative prediction of QCD is that there exist composite particles made solely of gluons called glueballs that have not yet been definitively observed experimentally. A definitive observation of a glueball with the properties predicted by QCD would strongly confirm the theory. In principle, if glueballs could be definitively ruled out, this would be a serious experimental blow to QCD. But, as of 2013, scientists are unable to confirm or deny the existence of glueballs definitively, despite the fact that particle accelerators have sufficient energy to generate them.

Cross-relations to Solid State Physics

There are unexpected cross-relations to solid state physics. For example, the notion of gauge invariance forms the basis of the well-known Mattis spin glasses, which are systems with the usual spin degrees of freedom $s_i = \pm 1$ for $i = 1,...,N$, with the special fixed "random" couplings $J_{i,k} = \epsilon_i J_0 \epsilon_k$. Here the ε_i and ε_k quantities can independently and "randomly" take the values ± 1, which corresponds to a most-simple gauge transformation $(s_i \to s_i \cdot \dot{\sigma}_i \quad J_{i,k} \to \dot{\sigma}_i J_{i,k} \dot{\sigma}_k \quad s_k \to s_k \cdot \dot{\sigma}_k)$. This means that thermodynamic expectation values of measurable quantities, e.g. of the energy $\mathcal{H} := -\sum_{i,k} s_i J_{i,k} s_k$, are invariant.

However, here the *coupling degrees of freedom* $J_{i,k}$, which in the QCD correspond to the *gluons*, are "frozen" to fixed values (quenching). In contrast, in the QCD they "fluctuate" (annealing), and through the large number of gauge degrees of freedom the entropy plays an important role.

For positive J_0 the thermodynamics of the Mattis spin glass corresponds in fact simply to a "ferromagnet in disguise", just because these systems have no "frustration" at all. This term is a basic measure in spin glass theory. Quantitatively it is identical with the loop product $P_W := J_{i,k} J_{k,l}...J_{n,m} J_{m,i}$ along a closed loop W. However, for a Mattis spin glass – in contrast to "genuine" spin glasses – the quantity P_W never becomes negative.

The basic notion "frustration" of the spin-glass is actually similar to the Wilson loop quantity of the QCD. The only difference is again that in the QCD one is dealing with SU(3) matrices, and that one is dealing with a "fluctuating" quantity. Energetically, perfect absence of frustration should be non-favorable and atypical for a spin glass, which means that one should add the loop product to the Hamiltonian, by some kind of term representing a "punishment". In the QCD the Wilson loop is essential for the Lagrangian rightaway.

The relation between the QCD and "disordered magnetic systems" (the spin glasses belong to them) were additionally stressed in a paper by Fradkin, Huberman and Shenker, which also stresses the notion of duality.

A further analogy consists in the already mentioned similarity to polymer physics, where, analogously to Wilson Loops, so-called "entangled nets" appear, which are important for the formation of the entropy-elasticity (force proportional to the length) of a rubber band. The non-abelian character of the SU(3) corresponds thereby to the non-trivial "chemical links", which glue different loop segments together, and "asymptotic freedom" means in the polymer analogy simply the fact that in the short-wave limit, i.e. for $0 \leftarrow \lambda_w \ll R_c$ (where R_c is a characteristic correlation length for the glued loops, corresponding to the above-mentioned "bag radius", while λ_w is the wavelength of an excitation) any non-trivial correlation vanishes totally, as if the system had crystallized.

There is also a correspondence between confinement in QCD – the fact that the color field is only different from zero in the interior of hadrons – and the behaviour of the usual magnetic field in the theory of type-II superconductors: there the magnetism is confined to the interiour of the Abrikosov flux-line lattice, i.e., the London penetration depth λ of that theory is analogous to the confinement radius R_c of quantum chromo-dynamics. Mathematically, this correspondendence is supported by the second term, $\propto g G_\mu^a \bar{\psi}_i \gamma^\mu T_{ij}^a \psi_j$, on the r.h.s. of the Lagrangian.

Quantum Field Theory

In theoretical physics, quantum field theory (QFT) is the theoretical framework for constructing quantum mechanical models of subatomic particles in particle physics and quasiparticles in condensed matter physics. A QFT treats particles as excited states of the underlying physical field, so these are called field quanta.

In quantum field theory, quantum mechanical interactions between particles are described by interaction terms between the corresponding underlying quantum fields. These interactions are conveniently visualized by Feynman diagrams, that also serve as a formal tool to evaluate various processes.

History

Historically, the development began in the 1920s with the quantization of the electromagnetic field, the quantization being based on an analogy of the eigenmode expansion of a vibrating string with fixed endpoints. In Weinberg (2005), QFT is brought forward as an unavoidable consequence of the reconciliation of quantum mechanics with special relativity.

Early Development

Max Born (1882–1970), one of the founders of quantum field theory.

He is also known for the Born rule that introduced the probabilistic interpretation in quantum mechanics. He received the 1954 Nobel Prize in Physics together with Walther Bothe.

The first achievement of quantum field theory, namely quantum electrodynamics (QED), is "still the paradigmatic example of a successful quantum field theory" according to Weinberg (2005). Ordinarily, QM cannot give an account of photons which constitute the prime case of relativistic 'particles'. Since photons have rest mass zero, and correspondingly travel in the vacuum at the speed c, a non-relativistic theory such as ordinary QM cannot give even an approximate description. Photons are implicit in the emission and absorption processes which have to be postulated; for instance, when one of an atom's electrons makes a transition between energy levels. The formalism of QFT is needed for an explicit description of photons. In fact most topics in the early development of quantum theory (the so-called old quantum theory, 1900–25) were related to the interaction of radiation and matter and thus should be treated by quantum field theoretical methods. However, quantum mechanics as formulated by Dirac, Heisenberg, and Schrödinger in 1926–27 started from atomic spectra and did not focus much on problems of radiation.

As soon as the conceptual framework of quantum mechanics was developed, a small group of theoreticians tried to extend quantum methods to electromagnetic fields. A good example is the famous paper by Born, Jordan & Heisenberg (1926). P. Jordan was especially acquainted with the literature on light quanta and made important contributions to QFT. The basic idea was that in QFT the electromagnetic field should be represented by matrices in the same way that position and momentum were represented in QM by matrices in matrix mechanics. The ideas of QM were extended to systems having an infinite number of degrees of freedom.

The inception of QFT is usually considered to be Dirac's famous 1927 paper on "The quantum theory of the emission and absorption of radiation". Here Dirac coined the name "quantum electrodynamics" (QED) for the part of QFT that was developed first. Dirac supplied a systematic procedure for transferring the characteristic quantum phenomenon of discreteness of physical quantities from the quantum-mechanical treatment of particles to a corresponding treatment of fields. Employing the theory of the quantum harmonic oscillator, Dirac gave a theoretical description of how photons appear in the quantization of the electromagnetic radiation field. Later, Dirac's procedure became a model for the quantization of other fields as well. These first approaches to QFT were further developed during the following three years. P. Jordan introduced creation and annihilation operators for fields obeying Fermi–Dirac statistics. These differ from the corresponding operators for Bose–Einstein statistics in that the former satisfy *anti-commutation relations* while the latter satisfy commutation relations.

The methods of QFT could be applied to derive equations resulting from the quantum-mechanical (field-like) treatment of particles, e.g. the Dirac equation, the Klein–

Gordon equation and the Maxwell equations. Schweber points out that the idea and procedure of second quantization goes back to Jordan, in a number of papers from 1927, while the expression itself was coined by Dirac. Some difficult problems concerning commutation relations, statistics, and Lorentz invariance were eventually solved. The first comprehensive account of a general theory of quantum fields, in particular, the method of canonical quantization, was presented by Heisenberg & Pauli in 1929. Whereas Jordan's second quantization procedure applied to the coefficients of the normal modes of the field, Heisenberg & Pauli started with the fields themselves and subjected them to the canonical procedure. Heisenberg and Pauli thus established the basic structure of QFT as presented in modern introductions to QFT. Fermi and Dirac, as well as Fock and Podolsky, presented different formulations which played a heuristic role in the following years.

Quantum electrodynamics rests on two pillars, see e.g., the short and lucid "Historical Introduction" of Scharf (2014). The first pillar is the quantization of the electromagnetic field, i.e., it is about photons as the quantized excitations or 'quanta' of the electromagnetic field. This procedure will be described in some more detail in the section on the particle interpretation. As Weinberg points out the "photon is the only particle that was known as a field before it was detected as a particle" so that it is natural that QED began with the analysis of the radiation field. The second pillar of QED consists of the relativistic theory of the electron, centered on the Dirac equation.

The Problem of Infinities

The Emergence of Infinities

Pascual Jordan (1902–1980), doctoral student of Max Born, was a pioneer in quantum field theory, coauthoring a number of seminal papers with Born and Heisenberg.

Jordan algebras were introduced by him to formalize the notion of an algebra of observables in quantum mechanics. He was awarded the Max Planck medal 1954.

Quantum field theory started with a theoretical framework that was built in analogy to quantum mechanics. Although there was no unique and fully developed theory, quantum field theoretical tools could be applied to concrete processes. Examples are the scattering of radiation by free electrons, Compton scattering, the collision between relativistic electrons or the production of electron-positron pairs by photons. Calculations to the first order of approximation were quite successful, but most people working in the field thought that QFT still had to undergo a major change. On the one side, some calculations of effects for cosmic rays clearly differed from measurements. On the other side and, from a theoretical point of view more threatening, calculations of higher orders of the perturbation series led to infinite results. The self-energy of the electron as well as vacuum fluctuations of the electromagnetic field seemed to be infinite. The perturbation expansions did not converge to a finite sum and even most individual terms were divergent.

The various forms of infinities suggested that the divergences were more than failures of specific calculations. Many physicists tried to avoid the divergences by formal tricks (truncating the integrals at some value of momentum, or even ignoring infinite terms) but such rules were not reliable, violated the requirements of relativity and were not considered as satisfactory. Others came up with first ideas of coping with infinities by a redefinition of the parameters of the theory and using a measured finite value, for example of the charge of the electron, instead of the infinite 'bare' value. This process is called renormalization.

From the point of view of the philosophy of science, it is remarkable that these divergences did not give enough reason to discard the theory. The years from 1930 to the beginning of World War II were characterized by a variety of attitudes towards QFT. Some physicists tried to circumvent the infinities by more-or-less arbitrary prescriptions, others worked on transformations and improvements of the theoretical framework. Most of the theoreticians believed that QED would break down at high energies. There was also a considerable number of proposals in favor of alternative approaches. These proposals included changes in the basic concepts e.g. negative probabilities and interactions at a distance instead of a field theoretical approach, and a methodological change to phenomenological methods that focusses on relations between observable quantities without an analysis of the microphysical details of the interaction, the so-called S-matrix theory where the basic elements are amplitudes for various scattering processes.

Despite the feeling that QFT was imperfect and lacking rigor, its methods were extended to new areas of applications. In 1933 Fermi's theory of the beta decay started with conceptions describing the emission and absorption of photons, transferred them to beta radiation and analyzed the creation and annihilation of electrons and neutrinos described by the weak interaction. Further applications of QFT outside of quantum electrodynamics succeeded in nuclear physics with the strong interaction. In 1934 Pauli & Weisskopf showed that a new type of fields (scalar fields), described by the Klein–

Gordon equation, could be quantized. This is another example of second quantization. This new theory for matter fields could be applied a decade later when new particles, pions, were detected.

The Taming of Infinities

Werner Heisenberg (1901–1976), doctoral student of Arnold Sommerfeld, was one of the founding fathers of quantum mechanics.

In particular, he introduced the version of quantum mechanics known as matrix mechanics, but is now more known for the Heisenberg uncertainty relations. He was awarded the Nobel prize in physics 1932 together with Erwin Schrödinger and Paul Dirac.

After the end of World War II more reliable and effective methods for dealing with infinities in QFT were developed, namely coherent and systematic rules for performing relativistic field theoretical calculations, and a general renormalization theory. On three famous conferences, the Shelter Island Conference 1947, the Pocono Conference 1948, and the 1949 Oldstone Conference, developments in theoretical physics were confronted with relevant new experimental results. In the late forties, there were two different ways to address the problem of divergences. One of these was discovered by Richard Feynman, the other one (based on an operator formalism) by Julian Schwinger and independently by Sin-Itiro Tomonaga. In 1949 Freeman Dyson showed that the two approaches are in fact equivalent. Thus, Freeman Dyson, Feynman, Schwinger, and Tomonaga became the inventors of renormalization theory. The most spectacular experimental successes of renormalization theory were the calculations of the anomalous magnetic moment of electron and the Lamb shift in the spectrum of hydrogen. These successes were so outstanding because the theoretical results were in better agreement with high precision experiments than anything in physics before. Nevertheless, mathematical problems lingered on and prompted a search for rigorous formulations.

The basic idea of renormalization is to avoid divergences that appear in physical predictions by shifting them into a part of the theory where they do not influence empirical propositions. Dyson could show that a rescaling of charge and mass ('renormalization') is sufficient to remove all divergences in QED to all orders of perturbation theory. In general, a QFT is called renormalizable, if all infinities can be absorbed into a redefinition of a finite number of coupling constants and masses. A consequence is that the physical charge and mass of the electron must be measured and cannot be computed from first principles. Perturbation theory gives well-defined predictions only in renormalizable quantum field theories, and luckily QED, the first fully developed QFT, belonged to this class of renormalizable theories. There are various technical procedures to renormalize a theory. One way is to cut off the integrals in the calculations at a certain value Λ of the momentum which is large but finite. This cut-off procedure is successful if, after taking the limit $\Lambda \to \infty$, the resulting quantities are independent of Λ. Part II of Peskin & Schroeder (1995) gives an extensive description of renormalization.

Richard Feynman (1918–1988)

His 1945 PhD thesis developed the path integral formulation of ordinary quantum mechanics. This was later generalized to field theory.

Feynman's formulation of QED is of special interest from a philosophical point of view. His so-called space-time approach is visualized by the famous Feynman diagrams that look like depicting paths of particles. Feynman's method of calculating scattering amplitudes is based on the functional integral formulation of field theory. A set of graphical rules can be derived so that the probability of a specific scattering process can be calculated by drawing a diagram of that process and then using the diagram to write down the mathematical expressions for calculating its amplitude. The diagrams provide an effective way to organize and visualize the various terms in the perturbation series, and they seem to display the flow of electrons and photons during the scattering process. External lines in the diagrams represent incoming and outgoing particles, internal lines are connected with virtual particles and vertices with interactions. Each of these graphical elements is associated with mathematical expressions that contribute to the amplitude of the respective process. The diagrams are part of Feynman's very efficient and elegant algorithm for computing the probability of scattering processes. The idea of particles traveling from one point to another was heuristically useful in constructing the theory. This heuristics, based on Huygen's principle, is useful for concrete calculations and actually give the correct particle propagators as derived more rigorously. Nevertheless, an analysis of the theoretical justification of the space-time approach shows that its success does not imply that particle paths have to be taken seriously. General arguments against a particle interpretation of QFT clearly exclude that the diagrams represent paths of particles in the interaction area. Feynman himself was not particularly interested in ontological questions.

Gauge Theory and the Standard Model

In the beginning of the 1950s, QED had become a reliable theory which no longer counted as preliminary. It took two decades from writing down the first equations until QFT could be applied to interesting physical problems in a systematic way. The new developments made it possible to apply QFT to new particles and new interactions. In the following decades QFT was extended to describe not only the electromagnetic force but also weak and strong interaction so that new Lagrangians had to be found which contain new classes of 'particles' or quantum fields. The research aimed at a more comprehensive theory of matter and in the end at a *unified theory of all interactions*.

New theoretical concepts had to be introduced, mainly connected with non-Abelian gauge theories (the effort of developing such theories started in 1954 with the work of Yang and Mills) and spontaneous symmetry breaking. Today there are trustworthy theories of the strong, weak, and electromagnetic interactions of elementary particles which have a similar structure as QED. A combined theory associated with the gauge group $SU(3) \times SU(2) \times U(1)$ is considered as the *standard model of elementary particle physics* which was achieved by Sheldon Glashow, Steven Weinberg and Abdul Salam in 1968, and Frank Wilczek, David Gross and David Politzer in 1973.

According to the standard model there are, on the one side, six types of leptons (e.g. the electron and its neutrino) and six types of quarks, where the members of both groups are all fermions with spin 1/2. On the other side, there are spin 1 particles (thus bosons) that mediate the interaction between elementary particles and the fundamental forces, namely the photon for electromagnetic interaction, two W and one Z-boson for weak interaction, and the gluon for strong interaction. Altogether there is good agreement with experimental data, for example, the masses of W+ and W− bosons (detected in 1983) confirmed the theoretical prediction within one percent deviation.

Common Trends in Particle, Condensed Matter and Statistical Physics

Renormalization Group Theory

Parallel developments in the understanding of phase transitions in condensed matter physics led to the study of the renormalization group. This involved the work of Leo Kadanoff (1966) and Michael Fisher (1973), which led to the seminal reformulation of quantum field theory by Kenneth G. Wilson in 1975.

Conformal Field Theory

During the same period, Kadanoff (1969) introduced an operator algebra formalism for the two-dimensional Ising model, a widely studied mathematical model of ferromagnetism in statistical physics. This development suggested that quantum field theory describes its scaling limit. Later, there developed the idea that a finite number of generating operators could represent all the correlation functions of the Ising model. In the 1980s, the existence of a much stronger symmetry for the scaling limit of two-dimensional critical systems was suggested by Alexander Belavin, Alexander Polyakov and Alexander Zamolodchikov in 1984, which eventually led to the development of conformal field theory, a special case of quantum field theory, which is presently employed successfully in different areas of particle physics and condensed matter physics.

Historiography

The first chapter in Weinberg (1995) is a very good short description of the earlier history of QFT. Detailed accounts of the historical development of QFT can be found, e.g., in Darrigol 1986, Schweber (1994) and Cao 1997a. Various historical and conceptual studies of the standard model are gathered in Hoddeson et al. 1997 and of renormalization theory in Brown 1993.

Definition

Quantum electrodynamics (QED) has one electron field and one photon field; quantum chromodynamics (QCD) has one field for each type of quark; and, in condensed matter, there is an atomic displacement field that gives rise to phonon particles. Edward Witten describes QFT as "by far" the most difficult theory in modern physics.

Dynamics

Ordinary quantum mechanical systems have a fixed number of particles, with each particle having a finite number of degrees of freedom. In contrast, the excited states of a quantum field can represent any number of particles. This makes quantum field theories especially useful for describing systems where the particle count/number may change over time, a crucial feature of relativistic dynamics.

States

QFT interaction terms are similar in spirit to those between charges with electric and magnetic fields in Maxwell's equations. However, unlike the classical fields of Maxwell's theory, fields in QFT generally exist in quantum superpositions of states and are subject to the laws of quantum mechanics.

Because the fields are continuous quantities over space, there exist excited states with arbitrarily large numbers of particles in them, providing QFT systems with effectively an infinite number of degrees of freedom. Infinite degrees of freedom can easily lead to divergences of calculated quantities (e.g., the quantities become infinite). Techniques such as renormalization of QFT parameters or discretization of spacetime, as in lattice QCD, are often used to avoid such infinities so as to yield physically plausible results.

Fields and Radiation

The gravitational field and the electromagnetic field are the only two fundamental fields in nature that have infinite range and a corresponding classical low-energy limit, which greatly diminishes and hides their "particle-like" excitations. Albert Einstein in 1905, attributed "particle-like" and discrete exchanges of momenta and energy, characteristic of "field quanta", to the electromagnetic field. Originally, his principal motivation was to explain the thermodynamics of radiation. Although the photoelectric effect and Compton scattering strongly suggest the existence of the photon, it might alternatively be explained by a mere quantization of emission; more definitive evidence of the quantum nature of radiation is now taken up into modern quantum optics as in the antibunching effect.

Varieties of Approaches

There is currently no complete quantum theory of the remaining fundamental force, gravity. Many of the proposed theories to describe gravity as a QFT postulate the existence of a graviton particle that mediates the gravitational force. Presumably, the as yet unknown correct quantum field-theoretic treatment of the gravitational field will behave like Einstein's general theory of relativity in the low-energy limit. Quantum field theory of the fundamental forces itself has been postulated to be the low-energy effective field theory limit of a more fundamental theory such as superstring theory.

Most theories in standard particle physics are formulated as *relativistic quantum field theories*, such as QED, QCD, and the Standard Model. QED, the quantum field-theoretic description of the electromagnetic field, approximately reproduces Maxwell's theory of electrodynamics in the low-energy limit, with small non-linear corrections to the Maxwell equations required due to virtual electron–positron pairs.

In the perturbative approach to quantum field theory, the full field interaction terms are approximated as a perturbative expansion in the number of particles involved. Each term in the expansion can be thought of as forces between particles being mediated by other particles. In QED, the electromagnetic force between two electrons is caused by an exchange of photons. Similarly, intermediate vector bosons mediate the weak force and gluons mediate the strong force in QCD. The notion of a force-mediating particle comes from perturbation theory, and does not make sense in the context of non-perturbative approaches to QFT, such as with bound states.

Principles

Classical and Quantum Fields

A classical field is a function defined over some region of space and time. Two physical phenomena which are described by classical fields are Newtonian gravitation, described by Newtonian gravitational field $g(x, t)$, and classical electromagnetism, described by the electric and magnetic fields $E(x, t)$ and $B(x, t)$. Because such fields can in principle take on distinct values at each point in space, they are said to have infinite degrees of freedom.

Classical field theory does not, however, account for the quantum-mechanical aspects of such physical phenomena. For instance, it is known from quantum mechanics that certain aspects of electromagnetism involve discrete particles—photons—rather than continuous fields. The business of *quantum* field theory is to write down a field that is, like a classical field, a function defined over space and time, but which also accommodates the observations of quantum mechanics. This is a *quantum field*.

It is not immediately clear *how* to write down such a quantum field, since quantum mechanics has a structure very unlike a field theory. In its most general formulation, quantum mechanics is a theory of abstract operators (observables) acting on an abstract state space (Hilbert space), where the observables represent physically observable quantities and the state space represents the possible states of the system under study. For instance, the fundamental observables associated with the motion of a single quantum mechanical particle are the position and momentum operators $\hat{x}$ and $\hat{p}$. Field theory, in contrast, treats x as a way to index the field rather than as an operator.

There are two common ways of developing a quantum field: the path integral formalism and canonical quantization. The latter of these is pursued in this article.

Lagrangian Formalism

Quantum field theory frequently makes use of the Lagrangian formalism from classical field theory. This formalism is analogous to the Lagrangian formalism used in classical mechanics to solve for the motion of a particle under the influence of a field. In classical field theory, one writes down a Lagrangian density, $\mathcal{L}$, involving a field, $\varphi(x,t)$, and possibly its first derivatives ($\partial\varphi/\partial t$ and $\nabla\varphi$), and then applies a field-theoretic form of the Euler–Lagrange equation. Writing coordinates $(t, x) = (x^0, x^1, x^2, x^3) = x^\mu$, this form of the Euler–Lagrange equation is

$$\frac{\partial}{\partial x^\mu}\left[\frac{\partial\mathcal{L}}{\partial(\partial\varphi/\partial x^\mu)}\right] - \frac{\partial\mathcal{L}}{\partial\varphi} = 0,$$

where a sum over μ is performed according to the rules of Einstein notation.

By solving this equation, one arrives at the "equations of motion" of the field. For example, if one begins with the Lagrangian density

$$\mathcal{L}(\varphi, \nabla\varphi) = -\rho(t, \mathbf{x})\varphi(t, \mathbf{x}) - \frac{1}{8\pi G}|\nabla\varphi|^2,$$

and then applies the Euler–Lagrange equation, one obtains the equation of motion

$$4\pi G\rho(t, \mathbf{x}) = \nabla^2\varphi.$$

This equation is Newton's law of universal gravitation, expressed in differential form in terms of the gravitational potential $\varphi(t, x)$ and the mass density $\rho(t, x)$. Despite the nomenclature, the "field" under study is the gravitational potential, φ, rather than the gravitational field, g. Similarly, when classical field theory is used to study electromagnetism, the "field" of interest is the electromagnetic four-potential (V/c, A), rather than the electric and magnetic fields E and B.

Quantum field theory uses this same Lagrangian procedure to determine the equations of motion for quantum fields. These equations of motion are then supplemented by commutation relations derived from the canonical quantization procedure described below, thereby incorporating quantum mechanical effects into the behavior of the field.

Single- and Many-particle Quantum Mechanics

In non-relativistic quantum mechanics, a particle (such as an electron or proton) is described by a complex wavefunction, $\psi(x, t)$, whose time-evolution is governed by the Schrödinger equation:

$$-\frac{\hbar^2}{2m}\frac{\partial^2}{\partial x^2}\psi(x,t) + V(x)\psi(x,t) = i\hbar\frac{\partial}{\partial t}\psi(x,t).$$

Here m is the particle's mass and $V(x)$ is the applied potential. Physical information about the behavior of the particle is extracted from the wavefunction by constructing expected values for various quantities; for example, the expected value of the particle's position is given by integrating $\psi^*(x)\, x\, \psi(x)$ over all space, and the expected value of the particle's momentum is found by integrating $-i\hbar\psi^*(x)d\psi/dx$. The quantity $\psi^*(x)$

$\psi(x)$ is itself in the Copenhagen interpretation of quantum mechanics interpreted as a probability density function. This treatment of quantum mechanics, where a particle's wavefunction evolves against a classical background potential $V(x)$, is sometimes called *first quantization*.

This description of quantum mechanics can be extended to describe the behavior of multiple particles, so long as the number and the type of particles remain fixed. The particles are described by a wavefunction $\psi(x_1, x_2, ..., x_N, t)$, which is governed by an extended version of the Schrödinger equation.

Often one is interested in the case where N particles are all of the same type (for example, the 18 electrons orbiting a neutral argon nucleus). As described in the article on identical particles, this implies that the state of the entire system must be either symmetric (bosons) or antisymmetric (fermions) when the coordinates of its constituent particles are exchanged. This is achieved by using a Slater determinant as the wavefunction of a fermionic system (and a Slater permanent for a bosonic system), which is equivalent to an element of the symmetric or antisymmetric subspace of a tensor product.

For example, the general quantum state of a system of N bosons is written as

$$|\phi_1 \cdots \phi_N\rangle = \sqrt{\frac{\prod_j N_j!}{N!}} \sum_{p \in S_N} |\phi_{p(1)}\rangle \otimes \cdots \otimes |\phi_{p(N)}\rangle,$$

where $|\phi_i\rangle$ are the single-particle states, N_j is the number of particles occupying state j, and the sum is taken over all possible permutations p acting on N elements. In general, this is a sum of $N!$ (N factorial) distinct terms. $\sqrt{\dfrac{\prod_j N_j!}{N!}}$ is a normalizing factor.

There are several shortcomings to the above description of quantum mechanics, which are addressed by quantum field theory. First, it is unclear how to extend quantum mechanics to include the effects of special relativity. Attempted replacements for the Schrödinger equation, such as the Klein–Gordon equation or the Dirac equation, have many unsatisfactory qualities; for instance, they possess energy eigenvalues that extend to $-\infty$, so that there seems to be no easy definition of a ground state. It turns out that such inconsistencies arise from relativistic wavefunctions not having a well-defined probabilistic interpretation in position space, as probability conservation is not a relativistically covariant concept. The second shortcoming, related to the first, is that in quantum mechanics there is no mechanism to describe particle creation and annihilation; this is crucial for describing phenomena such as pair production, which result from the conversion between mass and energy according to the relativistic relation $E = mc^2$.

Second Quantization

In this section, we will describe a method for constructing a quantum field theory called second quantization. This basically involves choosing a way to index the quantum mechanical degrees of freedom in the space of multiple identical-particle states. It is based on the Hamiltonian formulation of quantum mechanics.

Several other approaches exist, such as the Feynman path integral, which uses a Lagrangian formulation.

Bosons

For simplicity, we will first discuss second quantization for bosons, which form perfectly symmetric quantum states. Let us denote the mutually orthogonal single-particle states which are possible in the system by $|\phi_1\rangle, |\phi_2\rangle, |\phi_3\rangle$, and so on. For example, the 3-particle state with one particle in state $|\phi_1\rangle$ and two in state $|\phi_3\rangle$ is

$$\frac{1}{\sqrt{3}}\left[|\phi_1\rangle|\phi_2\rangle|\phi_2\rangle + |\phi_2\rangle|\phi_1\rangle|\phi_2\rangle + |\phi_2\rangle|\phi_2\rangle|\phi_1\rangle\right].$$

The first step in second quantization is to express such quantum states in terms of occupation numbers, by listing the number of particles occupying each of the single-particle states $|\phi_1\rangle, |\phi_2\rangle$, etc. This is simply another way of labelling the states. For instance, the above 3-particle state is denoted as

$$|1, 2, 0, 0, 0, \ldots\rangle.$$

An N-particle state belongs to a space of states describing systems of N particles. The next step is to combine the individual N-particle state spaces into an extended state space, known as Fock space, which can describe systems of any number of particles. This is composed of the state space of a system with no particles (the so-called vacuum state, written as $|0\rangle$), plus the state space of a 1-particle system, plus the state space of a 2-particle system, and so forth. States describing a definite number of particles are known as Fock states: a general element of Fock space will be a linear combination of Fock states. There is a one-to-one correspondence between the occupation number representation and valid boson states in the Fock space.

At this point, the quantum mechanical system has become a quantum field in the sense we described above. The field's elementary degrees of freedom are the occupation numbers, and each occupation number is indexed by a number j indicating which of the single-particle states $|\phi_1\rangle, |\phi_2\rangle, \ldots, |\phi_j\rangle, \ldots$ it refers to:

$$|N_1, N_2, N_3, \ldots, N_j, \ldots\rangle$$

The properties of this quantum field can be explored by defining creation and annihilation operators, which add and subtract particles. They are analogous to ladder operators in the quantum harmonic oscillator problem, which added and subtracted energy quanta. However, these operators literally create and annihilate particles of a given quantum state. The bosonic annihilation operator a_2 and creation operator $a_2^\dagger$ are easily defined in the occupation number representation as having the following effects:

$$a_2 \mid N_1, N_2, N_3, \ldots\rangle = \sqrt{N_2} \mid N_1, (N_2 - 1), N_3, \ldots\rangle,$$

$$a_2^\dagger \mid N_1, N_2, N_3, \ldots\rangle = \sqrt{N_2 + 1} \mid N_1, (N_2 + 1), N_3, \ldots\rangle.$$

It can be shown that these are operators in the usual quantum mechanical sense, i.e. linear operators acting on the Fock space. Furthermore, they are indeed Hermitian conjugates, which justifies the way we have written them. They can be shown to obey the commutation relation

$$\left[a_i, a_j\right] = 0 \quad , \quad \left[a_i^\dagger, a_j^\dagger\right] = 0 \quad , \quad \left[a_i, a_j^\dagger\right] = \delta_{ij},$$

where δ stands for the Kronecker delta. These are precisely the relations obeyed by the ladder operators for an infinite set of independent quantum harmonic oscillators, one for each single-particle state. Adding or removing bosons from each state is, therefore, analogous to exciting or de-exciting a quantum of energy in a harmonic oscillator.

Applying an annihilation operator a_k followed by its corresponding creation operator $a_k^\dagger$ returns the number N_k of particles in the kth single-particle eigenstate:

$$a_k^\dagger a_k \mid \ldots, N_k, \ldots = N_k \mid \ldots, N_k, \ldots$$

The combination of operators $a_k^\dagger a_k$ is known as the number operator for the kth eigenstate.

The Hamiltonian operator of the quantum field (which, through the Schrödinger equation, determines its dynamics) can be written in terms of creation and annihilation operators. For instance, for a field of free (non-interacting) bosons, the total energy of the field is found by summing the energies of the bosons in each energy eigenstate. If the kth single-particle energy eigenstate has energy E_k and there are N_k bosons in this state, then the total energy of these bosons is $E_k N_k$. The energy in the *entire* field is then a sum over k:

$$E_{\text{tot}} = \sum_k E_k N_k$$

This can be turned into the Hamiltonian operator of the field by replacing N_k with the corresponding number operator, $a_k^\dagger a_k$. This yields

$$H = \sum_k E_k a_k^\dagger a_k.$$

Fermions

It turns out that a different definition of creation and annihilation must be used for describing fermions. According to the Pauli exclusion principle, fermions cannot share quantum states, so their occupation numbers N_i can only take on the value 0 or 1. The fermionic annihilation operators c and creation operators † are defined by their actions on a Fock state thus

$$c_j \mid N_1, N_2, \ldots, N_j = 0, \ldots \rangle = 0$$

$$c_j \mid N_1, N_2, \ldots, N_j = 1, \ldots \rangle = (-1)^{(N_1 + \cdots + N_{j-1})} \mid N_1, N_2, \ldots, N_j = 0, \ldots \rangle$$

$$c_j^\dagger \mid N_1, N_2, \ldots, N_j = 0, \ldots \rangle = (-1)^{(N_1 + \cdots + N_{j-1})} \mid N_1, N_2, \ldots, N_j = 1, \ldots \rangle$$

$$c_j^\dagger \mid N_1, N_2, \ldots, N_j = 1, \ldots \rangle = 0.$$

These obey an anticommutation relation:

$$\{c_i, c_j\} = 0 \quad , \quad \{c_i^\dagger, c_j^\dagger\} = 0 \quad , \quad \{c_i, c_j^\dagger\} = \delta_{ij}.$$

One may notice from this that applying a fermionic creation operator twice gives zero, so it is impossible for the particles to share single-particle states, in accordance with the exclusion principle.

Field Operators

We have previously mentioned that there can be more than one way of indexing the degrees of freedom in a quantum field. Second quantization indexes the field by enumerating the single-particle quantum states. However, as we have discussed, it is more natural to think about a "field", such as the electromagnetic field, as a set of degrees of freedom indexed by position.

To this end, we can define *field operators* that create or destroy a particle at a particular point in space. In particle physics, these operators turn out to be more convenient to work with, because they make it easier to formulate theories that satisfy the demands of relativity.

Single-particle states are usually enumerated in terms of their momenta (as in the particle in a box problem.) We can construct field operators by applying the Fourier transform to the creation and annihilation operators for these states. For example, the bosonic field annihilation operator $\phi(\mathbf{r})$ is

$$\phi(\mathbf{r}) \stackrel{\text{def}}{=} \sum_j e^{i\mathbf{k}_j \cdot \mathbf{r}} a_j.$$

The bosonic field operators obey the commutation relation

$$\left[\phi(\mathbf{r}),\phi(\mathbf{r}')\right]=0 \quad,\quad \left[\phi^\dagger(\mathbf{r}),\phi^\dagger(\mathbf{r}')\right]=0 \quad,\quad \left[\phi(\mathbf{r}),\phi^\dagger(\mathbf{r}')\right]=\delta^3(\mathbf{r}-\mathbf{r}')$$

where $\delta(x)$ stands for the Dirac delta function. As before, the fermionic relations are the same, with the commutators replaced by anticommutators.

The field operator is not the same thing as a single-particle wavefunction. The former is an operator acting on the Fock space, and the latter is a quantum-mechanical amplitude for finding a particle in some position. However, they are closely related and are indeed commonly denoted with the same symbol. If we have a Hamiltonian with a space representation, say

$$H = -\frac{\hbar^2}{2m}\sum_i \nabla_i^2 + \sum_{i<j} U(|\mathbf{r}_i - \mathbf{r}_j|)$$

where the indices i and j run over all particles, then the field theory Hamiltonian (in the non-relativistic limit and for negligible self-interactions) is

$$H = -\frac{\hbar^2}{2m}\int d^3r\ \phi^\dagger(\mathbf{r})\nabla^2\phi(\mathbf{r}) + \frac{1}{2}\int d^3r \int d^3r'\ \phi^\dagger(\mathbf{r})\phi^\dagger(\mathbf{r}')U(|\mathbf{r}-\mathbf{r}'|)\phi(\mathbf{r}')\phi(\mathbf{r}).$$

This looks remarkably like an expression for the expectation value of the energy, with ϕ playing the role of the wavefunction. This relationship between the field operators and wave functions makes it very easy to formulate field theories starting from space projected Hamiltonians.

Dynamics

Once the Hamiltonian operator is obtained as part of the canonical quantization process, the time dependence of the state is described with the Schrödinger equation, just as with other quantum theories. Alternatively, the Heisenberg picture can be used where the time dependence is in the operators rather than in the states.

Implications

Unification of Fields and Particles

The "second quantization" procedure that we have outlined in the previous section takes a set of single-particle quantum states as a starting point. Sometimes, it is impossible to define such single-particle states, and one must proceed directly to quantum field theory. For example, a quantum theory of the electromagnetic field *must* be a quantum field theory, because it is impossible (for various reasons) to define a wavefunction for a single photon. In such situations, the quantum field theory can be constructed by examining the mechanical properties of the classical field and guessing the corresponding quantum theory. For free (non-interacting) quantum fields, the quantum field theories obtained in this way have the same properties as those obtained using second quantization, such as well-defined creation and annihilation operators obeying commutation or anticommutation relations.

Quantum field theory thus provides a unified framework for describing "field-like" objects (such as the electromagnetic field, whose excitations are photons) and "particle-like" objects (such as electrons, which are treated as excitations of an underlying electron field), so long as one can treat interactions as "perturbations" of free fields. There are still unsolved problems relating to the more general case of interacting fields that may or may not be adequately described by perturbation theory.

Physical Meaning of Particle Indistinguishability

The second quantization procedure relies crucially on the particles being identical. We would not have been able to construct a quantum field theory from a distinguishable many-particle system, because there would have been no way of separating and indexing the degrees of freedom.

Many physicists prefer to take the converse interpretation, which is that *quantum field theory explains what identical particles are.* In ordinary quantum mechanics, there is not much theoretical motivation for using symmetric (bosonic) or antisymmetric (fermionic) states, and the need for such states is simply regarded as an empirical fact. From the point of view of quantum field theory, particles are identical if and only if they are excitations of the same underlying quantum field. Thus, the question "why are all electrons identical?" arises from mistakenly regarding individual electrons as fundamental objects, when in fact it is only the electron field that is fundamental.

Particle Conservation and Non-conservation

During second quantization, we started with a Hamiltonian and state space describing a fixed number of particles (N), and ended with a Hamiltonian and state space for an arbitrary number of particles. Of course, in many common situations N is an important and perfectly well-defined quantity, e.g. if we are describing a gas of atoms sealed in a box. From the point of view of quantum field theory, such situations are described by quantum states that are eigenstates of the number operator , which measures the total number of particles present. As with any quantum mechanical observable, $\hat{N}$ is conserved if it commutes with the Hamiltonian. In that case, the quantum state is trapped in the N-particle subspace of the total Fock space, and the situation could equally well be described by ordinary N-particle quantum mechanics. (Strictly speaking, this is only true in the noninteracting case or in the low energy density limit of renormalized quantum field theories)

For example, we can see that the free boson Hamiltonian described above conserves particle number. Whenever the Hamiltonian operates on a state, each particle destroyed by an annihilation operator a_k is immediately put back by the creation operator $a_k^\dagger$..

On the other hand, it is possible, and indeed common, to encounter quantum states that are *not* eigenstates of $\hat{N}$, which do not have well-defined particle numbers. Such states are difficult or impossible to handle using ordinary quantum mechanics, but they can be easily described in quantum field theory as quantum superpositions of states having different values of N. For example, suppose we have a bosonic field whose particles can be created or destroyed by interactions with a fermionic field. The Hamiltonian of the combined system would be given by the Hamiltonians of the free boson and free fermion fields, plus a "potential energy" term such as

$$H_I = \sum_{k,q} V_q (a_q + a_{-q}^\dagger) c_{k+q}^\dagger c_k,$$

where $a_k^\dagger$ and a_k denotes the bosonic creation and annihilation operators, $c_k^\dagger$ and c_k denotes the fermionic creation and annihilation operators, and V_q is a parameter that describes the strength of the interaction. This "interaction term" describes processes in which a fermion in state k either absorbs or emits a boson, thereby being kicked into a different eigenstate $k+q$.. (In fact, this type of Hamiltonian is used to describe the interaction between conduction electrons and phonons in metals. The interaction between electrons and photons is treated in a similar way, but is a little more complicated because the role of spin must be taken into account.) One thing to notice here is that even if we start out with a fixed number of bosons, we will typically end up with a superposition of states with different numbers of bosons at later times. The number of fermions, however, is conserved in this case.

In condensed matter physics, states with ill-defined particle numbers are particularly important for describing the various superfluids. Many of the defining characteristics of a superfluid arise from the notion that its quantum state is a superposition of states with different particle numbers. In addition, the concept of a coherent state (used to model the laser and the BCS ground state) refers to a state with an ill-defined particle number but a well-defined phase.

Axiomatic Approaches

The preceding description of quantum field theory follows the spirit in which most physicists approach the subject. However, it is not mathematically rigorous. Over the past several decades, there have been many attempts to put quantum field theory on a firm mathematical footing by formulating a set of axioms for it. These attempts fall into two broad classes.

The first class of axioms, first proposed during the 1950s, include the Wightman, Oster-walder–Schrader, and Haag–Kastler systems. They attempted to formalize the physicists' notion of an "operator-valued field" within the context of functional analysis and enjoyed limited success. It was possible to prove that any quantum field theory satisfying these axioms satisfied certain general theorems, such as the spin-statistics theorem and the CPT theorem. Unfortunately, it proved extraordinarily difficult to show that any realistic field theory, including the Standard Model, satisfied these axioms. Most of the theories that could be treated with these analytic axioms were physically trivial, being restricted to low-dimensions and lacking interesting dynamics. The construction of theories satisfying one of these sets of axioms falls in the field of constructive quantum field theory. Important work was done in this area in the 1970s by Segal, Glimm, Jaffe and others.

During the 1980s, the second set of axioms based on geometric ideas was proposed. This line of investigation, which restricts its attention to a particular class of quantum field theories known as topological quantum field theories, is associated most closely with Michael Atiyah and Graeme Segal, and was notably expanded upon by Edward Witten, Richard Borcherds, and Maxim Kontsevich. However, most of the physically relevant quantum field theories, such as the Standard Model, are not topological quantum field theories; the quantum field theory of the fractional quantum Hall effect is a notable exception. The main impact of axiomatic topological quantum field theory has been on mathematics, with important applications in representation theory, algebraic topology, and differential geometry.

Finding the proper axioms for quantum field theory is still an open and difficult problem in mathematics. One of the Millennium Prize Problems—proving the existence of a mass gap in Yang–Mills theory—is linked to this issue.

Associated Phenomena

In the previous part of the article, we described the most general features of quantum field theories. Some of the quantum field theories studied in various fields of theoretical physics involve additional special ideas, such as renormalizability, gauge symmetry, and supersymmetry. These are described in the following sections.

Renormalization

Early in the history of quantum field theory, it was found that many seemingly innocuous calculations, such as the perturbative shift in the energy of an electron due to the presence of the electromagnetic field, give infinite results. The reason is that the perturbation theory for the shift in an energy involves a sum over all other energy levels, and there are infinitely many levels at short distances that each gives a finite contribution which results in a divergent series.

Many of these problems are related to failures in classical electrodynamics that were identified but unsolved in the 19th century, and they basically stem from the fact that many of the supposedly "intrinsic" properties of an electron are tied to the electromagnetic field that it carries around with it. The energy carried by a single electron—its self-energy—is not simply the bare value, but also includes the energy contained in its electromagnetic field, its attendant cloud of photons. The energy in a field of a spherical source diverges in both classical and quantum mechanics, but as discovered by Weisskopf with help from Furry, in quantum mechanics the divergence is much milder, going only as the logarithm of the radius of the sphere.

The solution to the problem, presciently suggested by Stueckelberg, independently by Bethe after the crucial experiment by Lamb, implemented at one loop by Schwinger, and systematically extended to all loops by Feynman and Dyson, with converging work by Tomonaga in isolated postwar Japan, comes from recognizing that all the infinities in the interactions of photons and electrons can be isolated into redefining a finite number of quantities in the equations by replacing them with the observed values: specifically the electron's mass and charge: this is called renormalization. The technique of renormalization recognizes that the problem is essentially purely mathematical, that extremely short distances are at fault. In order to define a theory on a continuum, first place a cutoff on the fields, by postulating that quanta cannot have energies above some extremely high value. This has the effect of replacing continuous space by a structure where very short wavelengths do not exist, as on a lattice. Lattices break rotational symmetry, and one of the crucial contributions made by Feynman, Pauli and Villars, and modernized by 't Hooft and Veltman, is a symmetry-preserving cutoff for perturbation theory (this process is called regularization). There is no known symmetrical cutoff outside of perturbation theory, so for rigorous or numerical work people often use an actual lattice.

On a lattice, every quantity is finite but depends on the spacing. When taking the limit of zero spacing, we make sure that the physically observable quantities like the observed electron mass stay fixed, which means that the constants in the Lagrangian defining the theory depend on the spacing. Hopefully, by allowing the constants to vary with the lattice spacing, all the results at long distances become insensitive to the lattice, defining a continuum limit.

The renormalization procedure only works for a certain class of quantum field theories, called *renormalizable quantum field theories*. A theory is *perturbatively renormalizable* when the constants in the Lagrangian only diverge at worst as logarithms of the lattice spacing for very short spacings. The continuum limit is then well defined in perturbation theory, and even if it is not fully well defined non-perturbatively, the problems only show up at distance scales that are exponentially small in the inverse coupling for weak couplings. The Standard Model of particle physics is perturbatively renormalizable, and so are its component theories (quantum electrodynamics/electroweak theory and quantum chromodynamics). Of the three components, quantum electrodynamics is believed to not have a continuum limit, while the asymptotically free SU(2) and SU(3) weak hypercharge and strong color interactions are nonperturbatively well defined.

The renormalization group describes how renormalizable theories emerge as the long distance low-energy effective field theory for any given high-energy theory. Because of this, renormalizable theories are insensitive to the precise nature of the underlying high-energy short-distance phenomena. This is a blessing because it allows physicists to formulate low energy theories without knowing the details of high energy phenomenon. It is also a curse, because once a renormalizable theory like the standard model is found to work, it gives very few clues to higher energy processes. The only way high energy processes can be seen in the standard model is when they allow otherwise forbidden events, or if they predict quantitative relations between the coupling constants.

Haag's Theorem

From a mathematically rigorous perspective, there exists no interaction picture in a Lorentz-covariant quantum field theory. This implies that the perturbative approach of Feynman diagrams in QFT is not strictly justified, despite producing vastly precise predictions validated by experiment. This is called Haag's theorem, but most particle physicists relying on QFT largely shrug it off.

Gauge Freedom

A gauge theory is a theory that admits a symmetry with a local parameter. For example, in every quantum theory the global phase of the wave function is arbitrary and does not represent something physical. Consequently, the theory is invariant under a global change of phases (adding a constant to the phase of all wave functions, everywhere); this is a global symmetry. In quantum electrodynamics, the theory is also invariant under a *local* change of phase, that is – one may shift the phase of all wave functions so that the shift may be different at every point in space-time. This is a *local* symmetry. However, in order for a well-defined derivative operator to exist, one must introduce a new field, the gauge field, which also transforms in order for the local change of variables (the phase in our example) not to affect the derivative. In quantum electrodynamics, this gauge field is the electromagnetic field. The change of local gauge of variables is termed gauge transformation. It is worth noting that by Noether's theorem, for every such symmetry there exists an associated conserved current. The aforementioned symmetry of the wavefunction under global phase changes implies the conservation of electric charge.

In quantum field theory the excitations of fields represent particles. The particle associated with excitations of the gauge field is the gauge boson, which is the photon in the case of quantum electrodynamics.

The degrees of freedom in quantum field theory are local fluctuations of the fields. The existence of a gauge symmetry reduces the number of degrees of freedom, simply because some fluctuations of the fields can be transformed to zero by gauge transformations, so they are equivalent to having no fluctuations at all, and they, therefore, have no physical meaning. Such fluctuations are usually called "non-physical degrees of freedom" or *gauge artifacts*; usually, some of them have a negative norm, making them inadequate for a consistent theory. Therefore, if a classical field theory has a gauge symmetry, then its quantized version (i.e. the corresponding quantum field theory) will have this symmetry as well. In other words, a gauge symmetry cannot have a quantum anomaly. If a gauge symmetry is anomalous (i.e. not kept in the quantum theory) then the theory is non-consistent: for example, in quantum electrodynamics, had there been a gauge anomaly, this would require the appearance of photons with longitudinal polarization and polarization in the time direction, the latter having a negative norm, rendering the theory inconsistent; another possibility would be for these photons to appear only in intermediate processes but not in the final products of any interaction, making the theory non-unitary and again inconsistent.

In general, the gauge transformations of a theory consist of several different transformations, which may not be commutative. These transformations are together described by a mathematical object known as a gauge group. Infinitesimal gauge transformations are the gauge group generators. Therefore, the number of gauge bosons is the group dimension (i.e. number of generators forming a basis).

All the fundamental interactions in nature are described by gauge theories. These are:

- Quantum chromodynamics, whose gauge group is SU(3). The gauge bosons are eight gluons.

- The electroweak theory, whose gauge group is U(1) × SU(2), (a direct product of U(1) and SU(2)).

- Gravity, whose classical theory is general relativity, admits the equivalence principle, which is a form of gauge symmetry. However, it is explicitly non-renormalizable.

Multivalued Gauge Transformations

The gauge transformations which leave the theory invariant involve, by definition, only single-valued gauge functions $\Lambda(x_i)$ which satisfy the Schwarz integrability criterion

$$\partial_{x_i x_j} \ddot{E} = \partial_{x_j x_i} \ddot{E}.$$

An interesting extension of gauge transformations arises if the gauge functions $\Lambda(x_i)$ are allowed to be multivalued functions which violate the integrability criterion. These are capable of changing the physical field strengths and are therefore not proper symmetry transformations. Nevertheless, the transformed field equations describe correctly the physical laws in the presence of the newly generated field strengths.

Supersymmetry

Supersymmetry assumes that every fundamental fermion has a superpartner that is a boson and vice versa. It was introduced in order to solve the so-called Hierarchy Problem, that is, to explain why particles not protected by any symmetry (like the Higgs boson) do not receive radiative corrections to its mass driving it to the larger scales (GUT, Planck...). It was soon realized that supersymmetry has other interesting properties: its gauged version is an extension of general relativity (Supergravity), and it is a key ingredient for the consistency of string theory.

The way supersymmetry protects the hierarchies is the following: since for every particle there is a superpartner with the same mass, any loop in a radiative correction is cancelled by the loop corresponding to its superpartner, rendering the theory UV finite.

Since no superpartners have yet been observed, if supersymmetry exists it must be broken (through a so-called soft term, which breaks supersymmetry without ruining its helpful features). The simplest models of this breaking require that the energy of the superpartners not be too high; in these cases, supersymmetry is expected to be observed by experiments at the Large Hadron Collider. The Higgs particle has been detected at the LHC, and no such superparticles have been discovered.

Gauge Theory

In physics, a gauge theory is a type of field theory in which the Lagrangian is invariant under a continuous group of local transformations.

The term *gauge* refers to redundant degrees of freedom in the Lagrangian. The transformations between possible gauges, called *gauge transformations*, form a Lie group— referred to as the *symmetry group* or the *gauge group* of the theory. Associated with any Lie group is the Lie algebra of group generators. For each group generator there necessarily arises a corresponding field (usually a vector field) called the *gauge field*. Gauge fields are included in the Lagrangian to ensure its invariance under the local group transformations (called *gauge invariance*). When such a theory is quantized, the quanta of the gauge fields are called *gauge bosons*. If the symmetry group is non-commutative, the gauge theory is referred to as *non-abelian*, the usual example being the Yang–Mills theory.

Many powerful theories in physics are described by Lagrangians that are invariant under some symmetry transformation groups. When they are invariant under a transformation identically performed at *every* point in the spacetime in which the physical processes occur, they are said to have a global symmetry. Local symmetry, the cornerstone of gauge theories, is a stricter constraint. In fact, a global symmetry is just a local symmetry whose group's parameters are fixed in spacetime.

Gauge theories are important as the successful field theories explaining the dynamics of elementary particles. Quantum electrodynamics is an abelian gauge theory with the symmetry group U(1) and has one gauge field, the electromagnetic four-potential, with the photon being the gauge boson. The Standard Model is a non-abelian gauge theory with the symmetry group U(1)×SU(2)×SU(3) and has a total of twelve gauge bosons: the photon, three weak bosons and eight gluons.

Gauge theories are also important in explaining gravitation in the theory of general relativity. Its case is somewhat unique in that the gauge field is a tensor, the Lanczos tensor. Theories of quantum gravity, beginning with gauge gravitation theory, also postulate the existence of a gauge boson known as the graviton. Gauge symmetries can be viewed as analogues of the principle of general covariance of general relativity in which the coordinate system can be chosen freely under arbitrary diffeomorphisms of space-time. Both gauge invariance and diffeomorphism invariance reflect a redundancy in the description of the system. An alternative theory of gravitation, gauge theory gravity, replaces the principle of general covariance with a true gauge principle with new gauge fields.

Historically, these ideas were first stated in the context of classical electromagnetism and later in general relativity. However, the modern importance of gauge symmetries appeared first in the relativistic quantum mechanics of electrons – quantum electrodynamics, elaborated on below. Today, gauge theories are useful in condensed matter, nuclear and high energy physics among other subfields.

History

The earliest field theory having a gauge symmetry was Maxwell's formulation, in 1864–65, of electrodynamics ("A Dynamical Theory of the Electromagnetic Field"). The importance of this symmetry remained unnoticed in the earliest formulations. Similarly unnoticed, Hilbert had derived the Einstein field equations by postulating the invariance of the action under a general coordinate transformation. Later Hermann Weyl, in an attempt to unify general relativity and electromagnetism, conjectured that *Eichinvarianz* or invariance under the change of scale (or "gauge") might also be a local symmetry of general relativity. After the development of quantum mechanics, Weyl, Vladimir Fock and Fritz London modified gauge by replacing the scale factor with a complex quantity and turned the scale transformation into a change of phase, which is a U(1) gauge symmetry. This explained the electromagnetic field effect on the wave function of a charged quantum mechanical particle. This was the first widely recognised gauge theory, popularised by Pauli in the 1940s.

In 1954, attempting to resolve some of the great confusion in elementary particle physics, Chen Ning Yang and Robert Mills introduced non-abelian gauge theories as models to understand the strong interaction holding together nucleons in atomic nuclei. (Ronald Shaw, working under Abdus Salam, independently introduced the same notion in his doctoral thesis.) Generalizing the gauge invariance of electromagnetism, they attempted to construct a theory based on the action of the (non-abelian) SU(2) symmetry group on the isospin doublet of protons and neutrons. This is similar to the action of the U(1) group on the spinor fields of quantum electrodynamics. In particle physics the emphasis was on using quantized gauge theories.

This idea later found application in the quantum field theory of the weak force, and its unification with electromagnetism in the electroweak theory. Gauge theories became even more attractive when it was realized that non-abelian gauge theories reproduced a feature called asymptotic freedom. Asymptotic freedom was believed to be an important characteristic of strong interactions. This motivated searching for a strong force gauge theory. This theory, now known as quantum chromodynamics, is a gauge theory with the action of the SU(3) group on the color triplet of quarks. The Standard Model unifies the description of electromagnetism, weak interactions and strong interactions in the language of gauge theory.

In the 1970s, Sir Michael Atiyah began studying the mathematics of solutions to the classical Yang–Mills equations. In 1983, Atiyah's student Simon Donaldson built on this work to show that the differentiable classification of smooth 4-manifolds is very different from their classification up to homeomorphism. Michael Freedman used Donaldson's work to exhibit exotic R⁴s, that is, exotic differentiable structures on Euclidean 4-dimensional space. This led to an increasing interest in gauge theory for its own sake, independent of its successes in fundamental physics. In 1994, Edward Witten and Nathan Seiberg invented gauge-theoretic techniques based on supersymmetry that enabled the calculation of certain topological invariants (the Seiberg–Witten invariants). These contributions to mathematics from gauge theory have led to a renewed interest in this area.

The importance of gauge theories in physics is exemplified in the tremendous success of the mathematical formalism in providing a unified framework to describe the quantum field theories of electromagnetism, the weak force and the strong force. This theory, known as the Standard Model, accurately describes experimental predictions regarding three of the four fundamental forces of nature, and is a gauge theory with the gauge group SU(3) × SU(2) × U(1). Modern theories like string theory, as well as general relativity, are, in one way or another, gauge theories.

Global and Local Symmetries

Global Symmetry

In physics, the mathematical description of any physical situation usually contains excess degrees of freedom; the same physical situation is equally well described by many equivalent mathematical configurations. For instance, in Newtonian dynamics, if two configurations are related by a Galilean transformation (an inertial change of reference frame) they represent the same physical situation. These transformations form a group of "symmetries" of the theory, and a physical situation corresponds not to an individual mathematical configuration but to a class of configurations related to one another by this symmetry group.

This idea can be generalized to include local as well as global symmetries, analogous to much more abstract "changes of coordinates" in a situation where there is no preferred "inertial" coordinate system that covers the entire physical system. A gauge theory is a mathematical model that has symmetries of this kind, together with a set of techniques for making physical predictions consistent with the symmetries of the model.

Example of Global Symmetry

When a quantity occurring in the mathematical configuration is not just a number but has some geometrical significance, such as a velocity or an axis of rotation, its representation as numbers arranged in a vector or matrix is also changed by a coordinate transformation. For instance, if one description of a pattern of fluid flow states that the fluid velocity in the neighborhood of (x=1, y=0) is 1 m/s in the positive x direction, then a description of the same situation in which the coordinate system has been rotated clockwise by 90 degrees states that the fluid velocity in the neighborhood of (x=0, y=1) is 1 m/s in the positive y direction. The coordinate transformation has affected both the coordinate system used to identify the *location* of the measurement and the basis in which its *value* is expressed. As long as this transformation is performed globally (affecting the coordinate basis in the same way at every point), the effect on values that represent the *rate of change* of some quantity along some path in space and time as it passes through point P is the same as the effect on values that are truly local to P.

Local Symmetry

Use of fiber bundles to describe local symmetries

In order to adequately describe physical situations in more complex theories, it is often necessary to introduce a "coordinate basis" for some of the objects of the theory that do not have this simple relationship to the coordinates used to label points in space and time. (In mathematical terms, the theory involves a fiber bundle in which the fiber at each point of the base space consists of possible coordinate bases for use when describing the values of objects at that point.) In order to spell out a mathematical configuration, one must choose a particular coordinate basis at each point (a *local section* of the fiber bundle) and express the values of the objects of the theory (usually "fields" in the physicist's sense) using this basis. Two such mathematical configurations are equivalent (describe the same physical situation) if they are related by a transformation of this abstract coordinate basis (a change of local section, or *gauge transformation*).

In most gauge theories, the set of possible transformations of the abstract gauge basis at an individual point in space and time is a finite-dimensional Lie group. The simplest such group is U(1), which appears in the modern formulation of quantum electrodynamics (QED) via its use of complex numbers. QED is generally regarded as the first, and simplest, physical gauge theory. The set of possible gauge transformations of the entire configuration of a given gauge theory also forms a group, the *gauge group* of the theory. An element of the gauge group can be parameterized by a smoothly varying function from the points of spacetime to the (finite-dimensional) Lie group, such that the value of the function and its derivatives at each point represents the action of the gauge transformation on the fiber over that point.

A gauge transformation with constant parameter at every point in space and time is analogous to a rigid rotation of the geometric coordinate system; it represents a global symmetry of the gauge representation. As in the case of a rigid rotation, this gauge transformation affects expressions that represent the rate of change along a path of some gauge-dependent quantity in the same way as those that represent a truly local quantity. A gauge transformation whose parameter is *not* a constant function is referred to as a local symmetry; its effect on expressions that involve a derivative is qualitatively different from that on expressions that don't. (This is analogous to a non-inertial change of reference frame, which can produce a Coriolis effect.)

Gauge Fields

The "gauge covariant" version of a gauge theory accounts for this effect by introducing a *gauge field* (in mathematical language, an Ehresmann connection) and formulating all rates of change in terms of the covariant derivative with respect to this connection. The gauge field becomes an essential part of the description of a mathematical configuration. A configuration in which the gauge field can be eliminated by a gauge transformation has the property that its field strength (in mathematical language, its curvature) is zero everywhere; a gauge theory is *not* limited to these configurations. In other words, the distinguishing characteristic of a gauge theory is that the gauge field does not merely compensate for a poor choice of coordinate system; there is generally no gauge transformation that makes the gauge field vanish.

When analyzing the dynamics of a gauge theory, the gauge field must be treated as a dynamical variable, similar to other objects in the description of a physical situation. In addition to its interaction with other objects via the covariant derivative, the gauge field typically contributes energy in the form of a "self-energy" term. One can obtain the equations for the gauge theory by:

- starting from a naïve ansatz without the gauge field (in which the derivatives appear in a "bare" form);

- listing those global symmetries of the theory that can be characterized by a continuous parameter (generally an abstract equivalent of a rotation angle);

- computing the correction terms that result from allowing the symmetry parameter to vary from place to place; and

- reinterpreting these correction terms as couplings to one or more gauge fields, and giving these fields appropriate self-energy terms and dynamical behavior.

This is the sense in which a gauge theory "extends" a global symmetry to a local symmetry, and closely resembles the historical development of the gauge theory of gravity known as general relativity.

Physical Experiments

Gauge theories are used to model the results of physical experiments, essentially by:

- limiting the universe of possible configurations to those consistent with the information used to set up the experiment, and then

- computing the probability distribution of the possible outcomes that the experiment is designed to measure.

The mathematical descriptions of the "setup information" and the "possible measurement outcomes" (loosely speaking, the "boundary conditions" of the experiment) are generally not expressible without reference to a particular coordinate system, including a choice of gauge. (If nothing else, one assumes that the experiment has been adequately isolated from "external" influence, which is itself a gauge-dependent statement.) Mishandling gauge dependence in boundary conditions is a frequent source of anomalies in gauge theory calculations, and gauge theories can be broadly classified by their approaches to anomaly avoidance.

Continuum Theories

The two gauge theories mentioned above (continuum electrodynamics and general relativity) are examples of continuum field theories. The techniques of calculation in a continuum theory implicitly assume that:

- given a completely fixed choice of gauge, the boundary conditions of an individual configuration can in principle be completely described;

- given a completely fixed gauge and a complete set of boundary conditions, the principle of least action determines a unique mathematical configuration (and therefore a unique physical situation) consistent with these bounds;

- the likelihood of possible measurement outcomes can be determined by:

    o   establishing a probability distribution over all physical situations determined by boundary conditions that are consistent with the setup information, o establishing a probability distribution of measurement outcomes for each possible physical situation, and

o convolving these two probability distributions to get a distribution of possible measurement outcomes consistent with the setup information; and

- fixing the gauge introduces no anomalies in the calculation, due either to gauge dependence in describing partial information about boundary conditions or to incompleteness of the theory.

These assumptions are close enough to be valid across a wide range of energy scales and experimental conditions, to allow these theories to make accurate predictions about almost all of the phenomena encountered in daily life, from light, heat, and electricity to eclipses and spaceflight. They fail only at the smallest and largest scales (due to omissions in the theories themselves) and when the mathematical techniques themselves break down (most notably in the case of turbulence and other chaotic phenomena).

Quantum Field Theories

Other than these classical continuum field theories, the most widely known gauge theories are quantum field theories, including quantum electrodynamics and the Standard Model of elementary particle physics. The starting point of a quantum field theory is much like that of its continuum analog: a gauge-covariant action integral that characterizes "allowable" physical situations according to the principle of least action. However, continuum and quantum theories differ significantly in how they handle the excess degrees of freedom represented by gauge transformations. Continuum theories, and most pedagogical treatments of the simplest quantum field theories, use a gauge fixing prescription to reduce the orbit of mathematical configurations that represent a given physical situation to a smaller orbit related by a smaller gauge group (the global symmetry group, or perhaps even the trivial group).

More sophisticated quantum field theories, in particular those that involve a non-abelian gauge group, break the gauge symmetry within the techniques of perturbation theory by introducing additional fields (the Faddeev–Popov ghosts) and counterterms motivated by anomaly cancellation, in an approach known as BRST quantization. While these concerns are in one sense highly technical, they are also closely related to the nature of measurement, the limits on knowledge of a physical situation, and the interactions between incompletely specified experimental conditions and incompletely understood physical theory. The mathematical techniques that have been developed in order to make gauge theories tractable have found many other applications, from solid-state physics and crystallography to low-dimensional topology.

Classical Gauge Theory

Classical Electromagnetism

Historically, the first example of gauge symmetry discovered was classical electromagnetism. In electrostatics, one can either discuss the electric field, E, or its corresponding electric potential, V. Knowledge of one makes it possible to find the other, except that potentials differing by a constant, $V \rightarrow V + C,$, correspond to the same electric field. This is because the electric field relates to *changes* in the potential from one point in space to another, and the constant C would cancel out when subtracting to find the change in potential. In terms of vector calculus, the electric field is the gradient of the potential, $\mathbf{E} = -\nabla V$. Generalizing from static electricity to electromagnetism, we have a second potential, the vector potential A, with

$$\mathbf{E} = -\nabla V - \frac{\partial \mathbf{A}}{\partial t}$$

$$\mathbf{B} = \nabla \times \mathbf{A}$$

The general gauge transformations now become not just $V \rightarrow V + C$ but

$$\mathbf{A} \rightarrow \mathbf{A} + \nabla f$$

$$V \rightarrow V - \frac{\partial f}{\partial t}$$

where f is any function that depends on position and time. The fields remain the same under the gauge transformation, and therefore Maxwell's equations are still satisfied. That is, Maxwell's equations have a gauge symmetry.

An example: Scalar O(n) gauge theory

> *The remainder of this section requires some familiarity with classical or quantum field theory, and the use of Lagrangians.*

> *Definitions in this section: gauge group, gauge field, interaction Lagrangian, gauge boson.*

The following illustrates how local gauge invariance can be "motivated" heuristically starting from global symmetry properties, and how it leads to an interaction between originally non-interacting fields.

Consider a set of n non-interacting real scalar fields, with equal masses m. This system is described by an action that is the sum of the (usual) action for each scalar field

$$S = \int d^4 x \sum_{i=1}^{n} \left[\frac{1}{2} \partial_\mu \varphi_i \partial^\mu \varphi_i - \frac{1}{2} m^2 \varphi_i^2 \right]$$

The Lagrangian (density) can be compactly written as

$$\mathcal{L} = \frac{1}{2}(\partial_\mu \Phi)^T \partial^\mu \Phi - \frac{1}{2}m^2 \Phi^T \Phi$$

by introducing a vector of fields

$$\Phi = (\varphi_1, \varphi_2, \ldots, \varphi_n)^T$$

The term ∂_μ is Einstein notation for the partial derivative of Φ in each of the four dimensions.

It is now transparent that the Lagrangian is invariant under the transformation

$$\Phi \mapsto \Phi' = G\Phi$$

whenever G is a *constant* matrix belonging to the n-by-n orthogonal group O(n). This is seen to preserve the Lagrangian, since the derivative of Φ transforms identically to Φ and both quantities appear inside dot products in the Lagrangian (orthogonal transformations preserve the dot product).

$$(\partial_\mu \Phi) \mapsto (\partial_\mu \Phi)' = G\partial_\mu \Phi$$

This characterizes the *global* symmetry of this particular Lagrangian, and the symmetry group is often called the gauge group; the mathematical term is structure group, especially in the theory of G-structures. Incidentally, Noether's theorem implies that invariance under this group of transformations leads to the conservation of the *currents*

$$J_\mu^a = i\partial_\mu \Phi^T T^a \Phi$$

where the T^a matrices are generators of the SO(n) group. There is one conserved current for every generator.

Now, demanding that this Lagrangian should have *local* O(n)-invariance requires that the G matrices (which were earlier constant) should be allowed to become functions of the space-time coordinates x.

In this case, the G matrices do not "pass through" the derivatives, when $G = G(x)$,

$$\partial_\mu(G\Phi) \neq G(\partial_\mu \Phi)$$

The failure of the derivative to commute with "G" introduces an additional term (in keeping with the product rule), which spoils the invariance of the Lagrangian. In order to rectify this we define a new derivative operator such that the derivative of Φ again transforms identically with Φ

$$(D_\mu \Phi)' = G D_\mu \Phi$$

This new "derivative" is called a (gauge) covariant derivative and takes the form

$$D_\mu = \partial_\mu + ig A_\mu$$

Where g is called the coupling constant; a quantity defining the strength of an interaction. After a simple calculation we can see that the gauge field $A(x)$ must transform as follows

$$A'_\mu = G A_\mu G^{-1} + \frac{i}{g}(\partial_\mu G) G^{-1}$$

The gauge field is an element of the Lie algebra, and can therefore be expanded as

$$A_\mu = \sum_a A_\mu^a T^a$$

There are therefore as many gauge fields as there are generators of the Lie algebra.

Finally, we now have a *locally gauge invariant* Lagrangian

$$\mathcal{L}_{\text{loc}} = \frac{1}{2}(D_\mu \Phi)^T D^\mu \Phi - \frac{1}{2} m^2 \Phi^T \Phi$$

Pauli uses the term *gauge transformation of the first type* to mean the transformation of Ö , while the compensating transformation in A is called a *gauge transformation of the second type.*

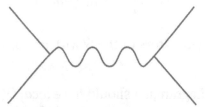

Feynman diagram of scalar bosons interacting via a gauge boson

The difference between this Lagrangian and the original *globally gauge-invariant* Lagrangian is seen to be the interaction Lagrangian

$$\mathcal{L}_{\text{int}} = i\frac{g}{2}\Phi^T A_\mu^T \partial^\mu \Phi + i\frac{g}{2}(\partial_\mu \Phi)^T A^\mu \Phi - \frac{g^2}{2}(A_\mu \Phi)^T A^\mu \Phi$$

This term introduces interactions between the n scalar fields just as a consequence of the demand for local gauge invariance. However, to make this interaction physical and not completely arbitrary, the mediator $A(x)$ needs to propagate in space. That is dealt with in the next section by adding yet another term, $\mathcal{L}_{gf}$, to the Lagrangian. In the quantized version of the obtained classical field theory, the quanta of the gauge field $A(x)$ are called gauge bosons. The interpretation of the interaction Lagrangian in quantum field theory is of scalar bosons interacting by the exchange of these gauge bosons.

The Yang–mills Lagrangian for the Gauge Field

The picture of a classical gauge theory developed in the previous section is almost complete, except for the fact that to define the covariant derivatives D, one needs to know the value of the gauge field $A(x)$ at all space-time points. Instead of manually specifying the values of this field, it can be given as the solution to a field equation. Further requiring that the Lagrangian that generates this field equation is locally gauge invariant as well, one possible form for the gauge field Lagrangian is (conventionally) written as

$$\mathcal{L}_{gf} = -\frac{1}{2}\mathrm{Tr}(F^{\mu\nu}F_{\mu\nu})$$

with

$$F_{\mu\nu} = \frac{1}{ig}[D_\mu, D_\nu]$$

and the trace being taken over the vector space of the fields. This is called the Yang–Mills action. Other gauge invariant actions also exist (e.g., nonlinear electrodynamics, Born–Infeld action, Chern–Simons model, theta term, etc.).

Note that in this Lagrangian term there is no field whose transformation counterweighs the one of A. Invariance of this term under gauge transformations is a particular case of *a priori* classical (geometrical) symmetry. This symmetry must be restricted in order to perform quantization, the procedure being denominated gauge fixing, but even after restriction, gauge transformations may be possible.

The complete Lagrangian for the gauge theory is now

$$\mathcal{L} = \mathcal{L}_{loc} + \mathcal{L}_{gf} = \mathcal{L}_{global} + \mathcal{L}_{int} + \mathcal{L}_{gf}$$

An example: Electrodynamics

As a simple application of the formalism developed in the previous sections, consider the case of electrodynamics, with only the electron field. The bare-bones action that generates the electron field's Dirac equation is

$$S = \int \bar{\psi}(i\hbar c \gamma^\mu \partial_\mu - mc^2)\psi \, \mathrm{d}^4 x$$

The global symmetry for this system is

$$\psi \mapsto e^{i\theta}\psi$$

The gauge group here is U(1), just rotations of the phase angle of the field, with the particular rotation determined by the constant θ.

"Localising" this symmetry implies the replacement of θ by $\theta(x)$. An appropriate covariant derivative is then

$$D_\mu = \partial_\mu - i\frac{e}{\hbar}A_\mu$$

Identifying the "charge" e with the usual electric charge (this is the origin of the usage of the term in gauge theories), and the gauge field $A(x)$ with the four-vector potential of electromagnetic field results in an interaction Lagrangian

$$\mathcal{L}_{\text{int}} = \frac{e}{\hbar}\bar{\psi}(x)\gamma^\mu\psi(x)A_\mu(x) = J^\mu(x)A_\mu(x)$$

where $J^\mu(x) = \frac{e}{\hbar}\bar{\psi}(x)\gamma^\mu\psi(x)$ is the electric current four vector in the Dirac field. The gauge principle is therefore seen to naturally introduce the so-called minimal coupling of the electromagnetic field to the electron field.

Adding a Lagrangian for the gauge field $A_\mu(x)$ in terms of the field strength tensor exactly as in electrodynamics, one obtains the Lagrangian used as the starting point in quantum electrodynamics.

$$\mathcal{L}_{\text{QED}} = \bar{\psi}(i\hbar c\gamma^\mu D_\mu - mc^2)\psi - \frac{1}{4\mu_0}F_{\mu\nu}F^{\mu\nu}$$

Mathematical Formalism

Gauge theories are usually discussed in the language of differential geometry. Mathematically, a *gauge* is just a choice of a (local) section of some principal bundle. A gauge transformation is just a transformation between two such sections.

Although gauge theory is dominated by the study of connections (primarily because it's mainly studied by high-energy physicists), the idea of a connection is not central to gauge theory in general. In fact, a result in general gauge theory shows that affine representations (i.e., affine modules) of the gauge transformations can be classified as sections of a jet bundle satisfying certain properties. There are representations that transform covariantly pointwise (called by physicists gauge transformations of the first kind), representations that transform as a connection form (called by physicists gauge transformations of the second kind, an affine representation)—and other more general representations, such as the B field in BF theory. There are more general nonlinear representations (realizations), but these are extremely complicated. Still, nonlinear sigma models transform nonlinearly, so there are applications.

If there is a principal bundle P whose base space is space or spacetime and structure group is a Lie group, then the sections of P form a principal homogeneous space of the group of gauge transformations.

Connections (gauge connection) define this principal bundle, yielding a covariant derivative ∇ in each associated vector bundle. If a local frame is chosen (a local basis of sections), then this covariant derivative is represented by the connection form A, a Lie algebra-valued 1-form, which is called the gauge potential in physics. This is evidently not an intrinsic but a frame-dependent quantity. The curvature form F, a Lie algebra-valued 2-form that is an intrinsic quantity, is constructed from a connection form by

$$\mathbf{F} = d\mathbf{A} + \mathbf{A} \wedge \mathbf{A}$$

where d stands for the exterior derivative and $\wedge$ stands for the wedge product. ($\mathbf{A}$ is an element of the vector space spanned by the generators T^a, and so the components of $\mathbf{A}$ do not commute with one another. Hence the wedge product $\mathbf{A} \wedge \mathbf{A}$ does not vanish.)

Infinitesimal gauge transformations form a Lie algebra, which is characterized by a smooth Lie-algebra-valued scalar, ε. Under such an infinitesimal gauge transformation,

$$\delta_\varepsilon \mathbf{A} = [\varepsilon, \mathbf{A}] - d\varepsilon$$

where $[\cdot, \cdot]$ is the Lie bracket.

One nice thing is that if $\delta_\varepsilon X = \varepsilon X$, then $\delta_\varepsilon DX = \varepsilon DX$ where D is the covariant derivative

$$DX \stackrel{\text{def}}{=} dX + \mathbf{A}X$$

Also, $\delta_\varepsilon \mathbf{F} = \varepsilon \mathbf{F}$, which means $\mathbf{F}$ transforms covariantly.

Not all gauge transformations can be generated by infinitesimal gauge transformations in general. An example is when the base manifold is a compact manifold without boundary such that the homotopy class of mappings from that manifold to the Lie group is nontrivial.

The *Yang–Mills action* is now given by

$$\frac{1}{4g^2} \int \mathrm{Tr}[*F \wedge F]$$

where * stands for the Hodge dual and the integral is defined as in differential geometry.

A quantity which is gauge-invariant (i.e., invariant under gauge transformations) is the Wilson loop, which is defined over any closed path, γ, as follows:

$$\chi^{(\rho)}\left(\mathcal{P}\left\{ e^{\int_\gamma A} \right\} \right)$$

where χ is the character of a complex representation ρ and $\mathcal{P}$ represents the path-ordered operator.

Quantization of Gauge Theories

Gauge theories may be quantized by specialization of methods which are applicable to any quantum field theory. However, because of the subtleties imposed by the gauge constraints there are many technical problems to be solved which do not arise in other field theories. At the same time, the richer structure of gauge theories allows simplification of some computations: for example Ward identities connect different renormalization constants.

Methods and Aims

The first gauge theory quantized was quantum electrodynamics (QED). The first methods developed for this involved gauge fixing and then applying canonical quantization. The Gupta–Bleuler method was also developed to handle this problem. Non-abelian gauge theories are now handled by a variety of means. Methods for quantization are covered in the article on quantization.

The main point to quantization is to be able to compute quantum amplitudes for various processes allowed by the theory. Technically, they reduce to the computations of certain correlation functions in the vacuum state. This involves a renormalization of the theory.

When the running coupling of the theory is small enough, then all required quantities may be computed in perturbation theory. Quantization schemes intended to simplify such computations (such as canonical quantization) may be called perturbative quantization schemes. At present some of these methods lead to the most precise experimental tests of gauge theories.

However, in most gauge theories, there are many interesting questions which are non-perturbative. Quantization schemes suited to these problems (such as lattice gauge theory) may be called non-perturbative quantization schemes. Precise computations in such schemes often require supercomputing, and are therefore less well-developed currently than other schemes.

Anomalies

Some of the symmetries of the classical theory are then seen not to hold in the quantum theory; a phenomenon called an anomaly. Among the most well known are:

- The scale anomaly, which gives rise to a *running coupling constant.* In QED this gives rise to the phenomenon of the Landau pole. In Quantum Chromodynamics (QCD) this leads to asymptotic freedom.

- The chiral anomaly in either chiral or vector field theories with fermions. This has close connection with topology through the notion of instantons. In QCD this anomaly causes the decay of a pion to two photons.

- The gauge anomaly, which must cancel in any consistent physical theory. In the electroweak theory this cancellation requires an equal number of quarks and leptons.

Pure Gauge

A pure gauge is the set of field configurations obtained by a gauge transformation on the null-field configuration, i.e., a gauge-transform of zero. So it is a particular "gauge orbit" in the field configuration's space.

Thus, in the abelian case, where $A_\mu(x) \to A'_\mu(x) = A_\mu(x) + \partial_\mu f(x)$, the pure gauge is just the set of field configurations $A'_\mu(x) = \partial_\mu f(x)$ for all $f(x)$.

Peccei–Quinn Theory

In particle physics, the Peccei–Quinn theory is a well-known proposal for the resolution of the strong CP problem. It was formulated by Roberto Peccei and Helen Quinn. The theory proposes that the QCD Lagrangian be extended with a CP-violating term known as the θ term. Because experiments have never measured a value for θ, its value must be small if it exists.

Peccei–Quinn theory predicts that the small θ parameter is explained by a dynamic field, rather than a constant value. Because particles arise within quantum fields, Peccei–Quinn theory predicts the existence of a new particle, the axion. The potential which this field carries causes it to have a value which naturally cancels, making the θ parameter uneventfully zero.

Peccei–Quinn symmetry presents θ as a functional component—a global U(1) symmetry under which a complex scalar field is charged. This symmetry is spontaneously broken by the vacuum expectation value obtained by this scalar field, and the axion is the massless Goldstone boson of this broken symmetry. If it is a gauge symmetry then the axion is superseded by the gauge boson, meaning that the gauge boson becomes mas-sive and the axion is no longer observable. This is phenome-nologically desirable because it leaves no massless particles, which are indeed not seen experimentally.

This Peccei–Quinn symmetry is inexact because it is anomalously broken by QCD instantons. If there is a compensatory term canceling the QCD anomaly breaking term, the axion becomes an exactly massless Goldstone boson and θ is no longer fixed. The effective potential of the axion is the summed potential above the QCD scale; with the potential term induced by nonperturbative QCD effects. If the axion is fundamental, or emerges at a scale far higher than the QCD scale, then the dimension 5 axion coupling term $a \, \mathrm{Tr}[\, F \, \wedge \, F \,]$ is suppressed by $1/\Lambda$ where Λ is the scale of the axion. Because of this, in order for θ to be so small at the minimum of the effective potential, the bare potential has to be many orders of magnitude smaller than the instanton induced potential, compounded by the Λ factor. This requires quite a bit of reconciliation with an approximate global symmetry, for which there is no current explanation.

References

- Pais, A. (1994) [1986]. Inward Bound: Of Matter and Forces in the Physical World (reprint ed.). Oxford, New York, Toronto: Oxford University Press. ISBN 978-0198519973.

- Schweber, S. S. (1994). QED and the Men Who Made It: Dyson, Feynman, Schwinger, and Tomonaga. Princeton University Press. ISBN 9780691033273.

- Peskin, M.; Schroeder, D. (1995). An Introduction to Quantum Field Theory. Westview Press. ISBN 0-201-50397-2.

- Scharf, Günter (2014) [1989]. Finite Quantum Electrodynamics: The Causal Approach (third ed.). Dover Publications. ISBN 978-0486492735.

- Zee, Anthony (2010). Quantum Field Theory in a Nutshell (2nd ed.). Princeton University Press. ISBN 978-0691140346.

4

Standard Models of Particle Physics

The theory of particle physics that concerns with the electromagnetic, weak and strong nuclear interactions is known as standard model of particle physics. The topics discusses in this chapter are Higgs mechanism, Cabibbo-Kobayashi-Maskawa matrix and spontaneous symmetry breaking. The major components of conceptual models are discussed in this chapter.

Standard Model

The Standard Model of particle physics is a theory concerning the electromagnetic, weak, and strong nuclear interactions, as well as classifying all the subatomic particles known. It was developed throughout the latter half of the 20th century, as a collaborative effort of scientists around the world. The current formulation was finalized in the mid-1970s upon experimental confirmation of the existence of quarks. Since then, discoveries of the top quark (1995), the tau neutrino (2000), and the Higgs boson (2012) have given further credence to the Standard Model. Because of its success in explaining a wide variety of experimental results, the Standard Model is sometimes regarded as the "theory of almost everything".

Although the Standard Model is believed to be theoretically self-consistent and has demonstrated huge and continued successes in providing experimental predictions, it does leave some phenomena unexplained and it falls short of being a complete theory of fundamental interactions. It does not incorporate the full theory of gravitation as described by general relativity, or account for the accelerating expansion of the universe (as possibly described by dark energy). The model does not contain any viable dark matter particle that possesses all of the required properties deduced from observational cosmology. It also does not incorporate neutrino oscillations (and their non-zero masses).

The development of the Standard Model was driven by theoretical and experimental particle physicists alike. For theorists, the Standard Model is a paradigm of a quantum field theory, which exhibits a wide range of physics including spontaneous symmetry breaking, anomalies and non-perturbative behavior. It is used as a basis for building more exotic models that incorporate hypothetical particles, extra dimensions, and elaborate symmetries (such as supersymmetry) in an attempt to explain experimental results at variance with the Standard Model, such as the existence of dark matter and neutrino oscillations.

Historical Background

The first step towards the Standard Model was Sheldon Glashow's discovery in 1961 of a way to combine the electromagnetic and weak interactions. In 1967 Steven Weinberg and Abdus Salam incorporated the Higgs mechanism into Glashow's electroweak interaction, giving it its modern form.

The Higgs mechanism is believed to give rise to the masses of all the elementary particles in the Standard Model. This includes the masses of the W and Z bosons, and the masses of the fermions, i.e. the quarks and leptons.

After the neutral weak currents caused by Z boson exchange were discovered at CERN in 1973, the electroweak theory became widely accepted and Glashow, Salam, and Weinberg shared the 1979 Nobel Prize in Physics for discovering it. The $W^{\pm}$ and Z° bosons were discovered experimentally in 1983; and the ratio of their masses was found to be as the Standard Model predicted.

The theory of the strong interaction (i.e. QCD), to which many contributed, acquired its modern form around 1973–74, when experiments confirmed that the hadrons were composed of fractionally charged quarks.

Overview

At present, matter and energy are best understood in terms of the kinematics and interactions of elementary particles. To date, physics has reduced the laws governing the behavior and interaction of all known forms of matter and energy to a small set of fundamental laws and theories. A major goal of physics is to find the "common ground" that would unite all of these theories into one integrated theory of everything, of which all the other known laws would be special cases, and from which the behavior of all matter and energy could be derived (at least in principle).

Particle Content

The Standard Model includes members of several classes of elementary particles (fermions, gauge bosons, and the Higgs boson), which in turn can be distinguished by other characteristics, such as color charge.

Fermions

The Standard Model includes 12 elementary particles of spin $\frac{1}{2}$ known as fermions. According to the spin-statistics theorem, fermions respect the Pauli exclusion principle. Each fermion has a corresponding antiparticle.

The fermions of the Standard Model are classified according to how they interact (or equivalently, by what charges they carry). There are six quarks (up, down, charm, strange, top, bottom), and six leptons (electron, electron neutrino, muon, muon neutrino, tau, tau neutrino). Pairs from each classification are grouped together to form a generation, with corresponding particles exhibiting similar physical behavior.

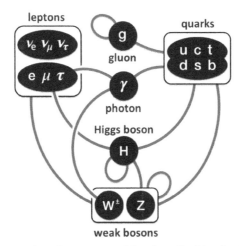

Summary of interactions between particles described by the Standard Model.

The defining property of the quarks is that they carry color charge, and hence, interact via the strong interaction. A phenomenon called color confinement results in quarks being very strongly bound to one another, forming color-neutral composite particles (hadrons) containing either a quark and an antiquark (mesons) or three quarks (baryons). The familiar proton and the neutron are the two baryons having the smallest mass. Quarks also carry electric charge and weak isospin. Hence, they interact with other fermions both electromagnetically and via the weak interaction.

The remaining six fermions do not carry colour charge and are called leptons. The three neutrinos do not carry electric charge either, so their motion is directly influenced only by the weak nuclear force, which makes them notoriously difficult to detect. However, by virtue of carrying an electric charge, the electron, muon, and tau all interact electromagnetically.

Each member of a generation has greater mass than the corresponding particles of lower generations. The first generation charged particles do not decay; hence all ordinary (baryonic) matter is made of such particles. Specifically, all atoms consist of electrons orbiting around atomic nuclei, ultimately constituted of up and down quarks. Second and third generation charged particles, on the other hand, decay with very short half lives, and are observed only in very high-energy environments. Neutrinos of all generations also do not decay, and pervade the universe, but rarely interact with baryonic matter.

Gauge Bosons

In the Standard Model, gauge bosons are defined as force carriers that mediate the strong, weak, and electromagnetic fundamental interactions.

Standard Model Interactions
(Forces Mediated by Gauge Bosons)

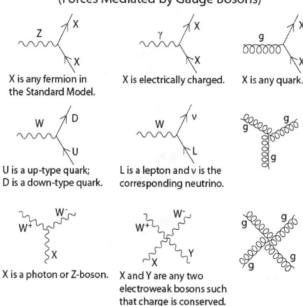

The above interactions form the basis of the standard model. Feynman diagrams in the standard model are built from these vertices. Modifications involving Higgs boson interactions and neutrino oscillations are omitted. The charge of the W bosons is dictated by the fermions they interact with; the conjugate of each listed vertex (i.e. reversing the direction of arrows) is also allowed.

Interactions in physics are the ways that particles influence other particles. At a macroscopic level, electromagnetism allows particles to interact with one another via electric and magnetic fields, and gravitation allows particles with mass to attract one another in accordance with Einstein's theory of general relativity. The Standard Model explains such forces as resulting from matter particles exchanging other particles, generally referred to as *force mediating particles*. When a force-mediating particle is exchanged, at a macroscopic level the effect is equivalent to a force influencing both of them, and the particle is therefore said to have *mediated* (i.e., been the agent of) that force. The Feynman diagram calculations, which are a graphical representation of the perturbation theory approximation, invoke "force mediating particles", and when applied to analyze high-energy scattering experiments are in reasonable agreement with the data. However, perturbation theory (and with it the concept of a "force-mediating particle") fails in other situations. These include low-energy quantum chromodynamics, bound states, and solitons.

The gauge bosons of the Standard Model all have spin (as do matter particles). The value of the spin is 1, making them bosons. As a result, they do not follow the Pauli exclusion principle that constrains fermions: thus bosons (e.g. photons) do not have a theoretical limit on their spatial density (number per volume). The different types of gauge bosons are described below.

- Photons mediate the electromagnetic force between electrically charged particles. The photon is massless and is well-described by the theory of quantum electrodynamics.

- The W+, W−, and Z gauge bosons mediate the weak interactions between particles of different flavors (all quarks and leptons). They are massive, with the Z being more massive than the W±. The weak interactions involving the W± exclusively act on *left-handed* particles and *right-handed* antiparticles. Furthermore, the W± carries an electric charge of +1 and −1 and couples to the electromagnetic interaction. The electrically neutral Z boson interacts with both left-handed particles and antiparticles. These three gauge bosons along with the photons are grouped together, as collectively mediating the electroweak interaction.

- The eight gluons mediate the strong interactions between color charged particles (the quarks). Gluons are massless. The eightfold multiplicity of gluons is labeled by a combination of color and anticolor charge (e.g. red–antigreen). Because the gluons have an effective color charge, they can also interact among themselves. The gluons and their interactions are described by the theory of quantum chromodynamics.

The interactions between all the particles described by the Standard Model are summarized by the diagrams on the right of this section.

Higgs Boson

The Higgs particle is a massive scalar elementary particle theorized by Peter Higgs in 1964 (see 1964 PRL symmetry breaking papers) and is a key building block in the Standard Model. It has no intrinsic spin, and for that reason is classified as a boson (like the gauge bosons, which have integer spin).

The Higgs boson plays a unique role in the Standard Model, by explaining why the other elementary particles, except the photon and gluon, are massive. In particular, the Higgs boson explains why the photon has no mass, while the W and Z bosons are very heavy. Elementary particle masses, and the differences between electromagnetism (mediated by the photon) and the weak force (mediated by the W and Z bosons), are critical to many aspects of the structure of microscopic (and hence macroscopic) matter. In electroweak theory, the Higgs boson generates the masses of the leptons (electron, muon, and tau) and quarks. As the Higgs boson is massive, it must interact with itself.

Because the Higgs boson is a very massive particle and also decays almost immediately when created, only a very high-energy particle accelerator can observe and record it. Experiments to confirm and determine the nature of the Higgs boson using the Large Hadron Collider (LHC) at CERN began in early 2010, and were performed at Fermilab's Tevatron until its closure in late 2011. Mathematical consistency of the Standard

Model requires that any mechanism capable of generating the masses of elementary particles becomes visible at energies above 1.4 TeV; therefore, the LHC (designed to collide two 7 to 8 TeV proton beams) was built to answer the question of whether the Higgs boson actually exists.

On 4 July 2012, the two main experiments at the LHC (ATLAS and CMS) both reported independently that they found a new particle with a mass of about 125 GeV/c^2 (about 133 proton masses, on the order of 10^{-25} kg), which is "consistent with the Higgs boson." Although it has several properties similar to the predicted "simplest" Higgs, they acknowledged that further work would be needed to conclude that it is indeed the Higgs boson, and exactly which version of the Standard Model Higgs is best supported if confirmed.

On 14 March 2013 the Higgs Boson was tentatively confirmed to exist.

Theoretical Aspects

Construction of the Standard Model Lagrangian

Parameters of the Standard Model			
Symbol	**Description**	**Renormalization scheme (point)**	**Value**
m_e	Electron mass		511 keV
m_μ	Muon mass		105.7 MeV
m_τ	Tau mass		1.78 GeV
m_u	Up quark mass	$\mu_{MS} = 2$ GeV	1.9 MeV
m_d	Down quark mass	$\mu_{MS} = 2$ GeV	4.4 MeV
m_s	Strange quark mass	$\mu_{MS} = 2$ GeV	87 MeV
m_c	Charm quark mass	$\mu_{MS} = m_c$	1.32 GeV
m_b	Bottom quark mass	$\mu_{MS} = m_b$	4.24 GeV
m_t	Top quark mass	On-shell scheme	172.7 GeV
θ_{12}	CKM 12-mixing angle		13.1°
θ_{23}	CKM 23-mixing angle		2.4°
θ_{13}	CKM 13-mixing angle		0.2°
δ	CKM CP-violating Phase		0.995
g_1 or g'	U(1) gauge coupling	$\mu_{MS} = m_z$	0.357
g_2 or g	SU(2) gauge coupling	$\mu_{MS} = m_z$	0.652
g_3 or g_s	SU(3) gauge coupling	$\mu_{MS} = m_z$	1.221
θ_{QCD}	QCD vacuum angle		~0
v	Higgs vacuum expectation value		246 GeV
m_H	Higgs mass		125.36±0.41 GeV (tentative)

Technically, quantum field theory provides the mathematical framework for the Standard Model, in which a Lagrangian controls the dynamics and kinematics of the theory. Each kind of particle is described in terms of a dynamical field that pervades spacetime. The construction of the Standard Model proceeds following the modern method of constructing most field theories: by first postulating a set of symmetries of the system, and then by writing down the most general renormalizable Lagrangian from its particle (field) content that observes these symmetries.

The global Poincaré symmetry is postulated for all relativistic quantum field theories. It consists of the familiar translational symmetry, rotational symmetry and the inertial reference frame invariance central to the theory of special relativity. The local SU(3)×SU(2)×U(1) gauge symmetry is an internal symmetry that essentially defines the Standard Model. Roughly, the three factors of the gauge symmetry give rise to the three fundamental interactions. The fields fall into different representations of the various symmetry groups of the Standard Model. Upon writing the most general La-grangian, one finds that the dynamics depend on 19 parameters, whose numerical val-ues are established by experiment. The parameters are summarized in the table above (note: the Higgs mass is at 125 GeV, the Higgs self-coupling strength $\lambda \sim \frac{1}{8}$).

Quantum Chromodynamics Sector

The quantum chromodynamics (QCD) sector defines the interactions between quarks and gluons, with SU(3) symmetry, generated by T^a. Since leptons do not interact with gluons, they are not affected by this sector. The Dirac Lagrangian of the quarks coupled to the gluon fields is given by

$Ga \mu$ is the SU(3) gauge field containing the gluons, γ^μ are the Dirac matrices, D and U are the Dirac spinors associated with up- and down-type quarks, and g_s is the strong coupling constant.

Electroweak Sector

The electroweak sector is a Yang–Mills gauge theory with the simple symmetry group U(1)×SU(2)$_L$,

$$\mathcal{L}_{EW} = \sum_\psi \bar{\psi}\gamma^\mu \left(i\partial_\mu - g'\frac{1}{2}Y_W B_\mu - g\frac{1}{2}\vec{\tau}_L \vec{W}_\mu \right)\psi,$$

where B_μ is the U(1) gauge field; Y_W is the weak hypercharge—the generator of the U(1) group; $\vec{W}_\mu$ is the three-component SU(2) gauge field; $\vec{\tau}_L$ are the Pauli matrices—infinitesimal generators of the SU(2) group (the subscript L indicates that they only act on left fermions) g' and g are coupling constants.

Higgs Sector

In the Standard Model, the Higgs field is a complex scalar of the group SU(2)$_L$:

$$\varphi = \frac{1}{\sqrt{2}}\begin{pmatrix}\varphi^{+} \\ \varphi^{0}\end{pmatrix},$$

where the indices + and 0 indicate the electric charge (Q) of the components. The weak isospin (Y_{W}) of both components is 1.

Before symmetry breaking, the Higgs Lagrangian is:

$$\mathcal{L}_{\mathrm{H}} = \varphi^{\dagger}\left(\partial^{\mu} - \frac{i}{2}\left(g'Y_{\mathrm{W}}B^{\mu} + g\vec{\tau}\vec{W}^{\mu}\right)\right)\left(\partial_{\mu} + \frac{i}{2}\left(g'Y_{\mathrm{W}}B_{\mu} + g\vec{\tau}\vec{W}_{\mu}\right)\right)\varphi - \frac{\lambda^{2}}{4}\left(\varphi^{\dagger}\varphi - v^{2}\right)^{2},$$

which can also be written as:

$$\mathcal{L}_{\mathrm{H}} = \left|\left(\partial_{\mu} + \frac{i}{2}\left(g'Y_{\mathrm{W}}B_{\mu} + g\vec{\tau}\vec{W}_{\mu}\right)\right)\varphi\right|^{2} - \frac{\lambda^{2}}{4}\left(\varphi^{\dagger}\varphi - v^{2}\right)^{2}.$$

Fundamental Forces

The Standard Model classified all four fundamental forces in nature. In the Standard Model, a force is described as an exchange of bosons between the objects affected, such as a photon for the electromagnetic force and a gluon for the strong interaction. Those particles are called force carriers.

The four fundamental forces of nature					
Property/Interaction	Gravitation	Weak	Electromagnetic	Strong	
		(Electroweak)		Fundamental	Residual
Acts on:	Mass - Energy	Flavor	Electric charge	Color charge	Atomic nuclei
Particles experiencing:	All	Quarks, leptons	Electrically charged	Quarks, Gluons	Hadrons
Particles mediating:	Graviton (theoretical)	W⁺ W⁻ Z⁰	γ	Gluons	Mesons
Strength in the scale of quarks:	10^{-41}	10^{-4}	1	60	Not applicable to quarks
Strength in the scale of protons/neutrons:	10^{-36}	10^{-7}	1	Not applicable to hadrons	20

Tests and Predictions

The Standard Model (SM) predicted the existence of the W and Z bosons, gluon, and the top and charm quarks before these particles were observed. Their predicted properties were experimentally confirmed with good precision.

The SM also makes several predictions about the decay of Z^0 bosons, which have been experimentally confirmed by the Large Electron-Positron Collider at CERN.

In May 2012 BaBar Collaboration reported that their recently analyzed data may suggest possible flaws in the Standard Model of particle physics. These data show that a particular type of particle decay called "B to D-star-tau-nu" happens more often than the Standard Model says it should. In this type of decay, a particle called the B-bar meson decays into a D meson, an antineutrino and a tau-lepton. While the level of certainty of the excess (3.4 sigma) is not enough to claim a break from the Standard Model, the results are a potential sign of something amiss and are likely to impact existing theories, including those attempting to deduce the properties of Higgs bosons.

On December 13, 2012, physicists reported the constancy, over space and time, of a basic physical constant of nature that supports the *standard model of physics*. The scientists, studying methanol molecules in a distant galaxy, found the change $(\Delta\mu/\mu)$ in the proton-to-electron mass ratio μ to be equal to "$(0.0 \pm 1.0) \times 10^{-7}$ at redshift z = 0.89" and consistent with "a null result".

Challenges

> **Unsolved problem in physics**:
>
> - What gives rise to the Standard Model of particle physics?
>
> - Why do particle masses and coupling constants have the values that we measure?
>
> - Why are there three generations of particles?
>
> - Why is there more matter than antimatter in the universe?
>
> - Where does Dark Matter fit into the model? Is it even a new particle?
>
> *(more unsolved problems in physics)*

Self-consistency of the Standard Model (currently formulated as a non-abelian gauge theory quantized through path-integrals) has not been mathematically proven. While regularized versions useful for approximate computations (for example lattice gauge theory) exist, it is not known whether they converge (in the sense of S-matrix elements) in the limit that the regulator is removed. A key question related to the consistency is the Yang–Mills existence and mass gap problem.

Experiments indicate that neutrinos have mass, which the classic Standard Model did not allow. To accommodate this finding, the classic Standard Model can be modified to include neutrino mass.

If one insists on using only Standard Model particles, this can be achieved by adding a non-renormalizable interaction of leptons with the Higgs boson. On a fundamental level, such an interaction emerges in the seesaw mechanism where heavy right-handed neutrinos are added to the theory. This is natural in the left-right symmetric extension of the Standard Model and in certain grand unified theories. As long as new physics appears below or around 10^{14} GeV, the neutrino masses can be of the right order of magnitude.

Theoretical and experimental research has attempted to extend the Standard Model into a Unified field theory or a Theory of everything, a complete theory explaining all physical phenomena including constants. Inadequacies of the Standard Model that motivate such research include:

- The model does not explain gravitation, although physical confirmation of a theoretical particle known as a graviton would account for it to a degree. Though it addresses strong and electroweak interactions, the Standard Model does not consistently explain the canonical theory of gravitation, general relativity, in terms of quantum field theory. The reason for this is, among other things, that quantum field theories of gravity generally break down before reaching the Planck scale. As a consequence, we have no reliable theory for the very early universe.

- Some physicists consider it to be *ad hoc* and inelegant, requiring 19 numerical constants whose values are unrelated and arbitrary. Although the Standard Model, as it now stands, can explain why neutrinos have masses, the specifics of neutrino mass are still unclear. It is believed that explaining neutrino mass will require an additional 7 or 8 constants, which are also arbitrary parameters.

- The Higgs mechanism gives rise to the hierarchy problem if some new physics (coupled to the Higgs) is present at high energy scales. In these cases, in order for the weak scale to be much smaller than the Planck scale, severe fine tuning of the parameters is required; there are, however, other scenarios that include quantum gravity in which such fine tuning can be avoided. There are also issues of Quantum triviality, which suggests that it may not be possible to create a consistent quantum field theory involving elementary scalar particles.

- The model is inconsistent with the emerging "Standard Model of cosmology." More common contentions include the absence of an explanation in the Standard Model for the observed amount of cold dark matter (CDM) and its contributions to dark energy, which are many orders of magnitude too large. It is also difficult to accommodate the observed predominance of matter over antimatter (matter/antimatter asymmetry). The isotropy and homogeneity of the visible universe over large distances seems to require a mechanism like cosmic inflation, which would also constitute an extension of the Standard Model.

- The existence of ultra-high-energy cosmic rays are difficult to explain under the Standard Model.

Currently, no proposed Theory of Everything has been widely accepted or verified.

Standard Model (Mathematical Formulation)

This article describes the mathematics of the Standard Model of particle physics, a gauge quantum field theory containing the internal symmetries of the unitary product group $SU(3) \times SU(2) \times U(1)$. The theory is commonly viewed as containing the fundamental set of particles – the leptons, quarks, gauge bosons and the Higgs particle.

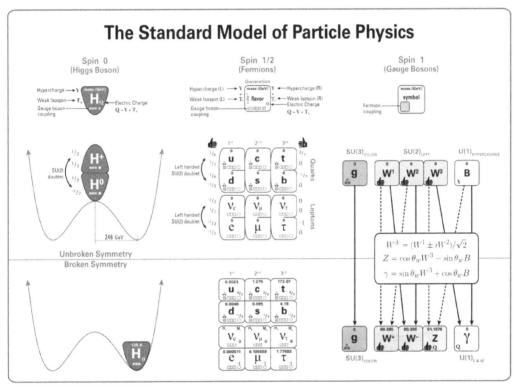

Standard Model of Particle Physics. The diagram shows the elementary particles of the Standard Model (the Higgs boson, the three generations of quarks and leptons, and the gauge bosons), including their names, masses, spins, charges, chiralities, and interactions with the strong, weak and electromagnetic forces. It also depicts the crucial role of the Higgs boson in electroweak symmetry breaking, and shows how the properties of the various particles differ in the (high-energy) symmetric phase (top) and the (low-energy) broken-symmetry phase (bottom).

The Standard Model is renormalizable and mathematically self-consistent, however despite having huge and continued successes in providing experimental predictions it does leave some unexplained phenomena. In particular, although the physics of special relativity is incorporated, general relativity is not, and the Standard Model will fail at energies or distances where the graviton is expected to emerge. Therefore, in a modern field theory context, it is seen as an effective field theory.

This article requires some background in physics and mathematics, but is designed as both an introduction and a reference.

Quantum Field Theory

The standard model is a quantum field theory, meaning its fundamental objects are *quantum fields* which are defined at all points in spacetime. These fields are

- the fermion fields, ψ, which account for "matter particles";

- the electroweak boson fields W_1, W_2, W_3, and B;

- the gluon field, G_a; and

- the Higgs field, φ.

That these are *quantum* rather than *classical* fields has the mathematical consequence that they are operator-valued. In particular, values of the fields generally do not commute. As operators, they act upon the quantum state (ket vector).

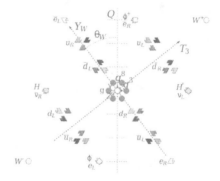

The pattern of weak isospin T_3, weak hypercharge Y_W, and color charge of all known elementary particles, rotated by the weak mixing angle to show electric charge Q, roughly along the vertical. The neutral Higgs field (gray square) breaks the electroweak symmetry and interacts with other particles to give them mass.

The dynamics of the quantum state and the fundamental fields are determined by the Lagrangian density $\mathcal{L}$ (usually for short just called the Lagrangian). This plays a role similar to that of the Schrödinger equation in non-relativistic quantum mechanics, but a Lagrangian is not an equation of motion – rather, it is a polynomial function of the fields and their derivatives, and used with the principle of least action. While it would be possible to derive a system of differential equations governing the fields from the Langrangian, it is more common to use other techniques to compute with quantum field theories.

The standard model is furthermore a gauge theory, which means there are degrees of freedom in the mathematical formalism which do not correspond to changes in the physical state. The gauge group of the standard model is $SU(3) \times SU(2) \times U(1)$, where $U(1)$ acts on B and φ, $SU(2)$ acts on W and φ, and $SU(3)$ acts on G. The fermion field ψ also transforms under these symmetries, although all of them leave some parts of it unchanged.

The Role of the Quantum Fields

In classical mechanics, the state of a system can usually be captured by a small set of variables, and the dynamics of the system is thus determined by the time evolution of these variables. In classical field theory, the *field* is part of the state of the system, so in order to describe it completely one effectively introduces separate variables for every point in spacetime (even though there are many restrictions on how the values of the field "variables" may vary from point to point, for example in the form of field equations involving partial derivatives of the fields).

In quantum mechanics, the classical variables are turned into operators, but these do not capture the state of the system, which is instead encoded into a wavefunction ψ or more abstract ket vector. If ψ is an eigenstate with respect to an operator P, then $P\psi = \lambda\psi$ for the corresponding eigenvalue λ, and hence letting an operator P act on ψ is analogous to multiplying ψ by the value of the classical variable to which P corresponds. By extension, a classical formula where all variables have been replaced by the corresponding operators will behave like an operator which, when it acts upon the state of the system, multiplies it by the analogue of the quantity that the classical formula would compute. The formula as such does however not contain any information about the state of the system; it would evaluate to the same operator regardless of what state the system is in.

Quantum fields relate to quantum mechanics as classical fields do to classical mechanics, i.e., there is a separate operator for every point in spacetime, and these operators do not carry any information about the state of the system; they are merely used to exhibit some aspect of the state, at the point to which they belong. In particular, the quantum fields are *not* wavefunctions, even though the equations which govern their time evolution may be deceptively similar to those of the corresponding wavefunction in a semiclassical formulation. There is no variation in strength of the fields between different points in spacetime; the variation that happens is rather one of phase factors.

Vectors, Scalars, and Spinors

Mathematically it may look as though all of the fields are vector-valued (in addition to being operator-valued), since they all have several components, can be multiplied by matrices, etc., but physicists assign a more specific physical meaning to the word: a vector is something which transforms like a four-vector under Lorentz transformations, and a scalar is something which is invariant under Lorentz transformations. The B, W_j, and G_a fields are all vectors in this sense, so the corresponding particles are said to be vector bosons. The Higgs field φ is a scalar.

The fermion field ψ does transform under Lorentz transformations, but not like a vector should; rotations will only turn it by half the angle a proper vector should. Therefore, these constitute a third kind of quantity, which is known as a spinor.

It is common to make use of abstract index notation for the vector fields, in which case the vector fields all come with a Lorentzian index μ, like so: B^μ, W_j^μ, and G_a^μ. If abstract index notation is used also for spinors then these will carry a spinorial index and the Dirac gamma will carry one Lorentzian and two spinorian indices, but it is more common to regard spinors as column matrices and the Dirac gamma γ_μ as a matrix which additionally carries a Lorentzian index. The Feynman slash notation can be used to turn a vector field into a linear operator on spinors, like so: $B = \gamma^\mu B_\mu$; this may involve raising and lowering indices.

Alternative Presentations of the Fields

As is common in quantum theory, there is more than one way to look at things. At first the basic fields given above may not seem to correspond well with the "fundamental particles" in the chart above, but there are several alternative presentations which, in particular contexts, may be more appropriate than those that are given above.

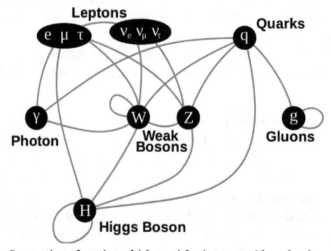

Connections denoting which particles interact with each other.

Fermions

Rather than having one fermion field ψ, it can be split up into separate components for each type of particle. This mirrors the historical evolution of quantum field theory, since the electron component ψ_e (describing the electron and its antiparticle the positron) is then the original ψ field of quantum electrodynamics, which was later accompanied by ψ_μ and ψ_τ fields for the muon and tauon respectively (and their antiparticles). Electroweak theory added $\psi_{\nu_e}, \psi_{\nu_\mu}$, and ψ_{ν_τ} for the corresponding neutrinos, and the quarks add still further components. In order to be four-spinors like the electron and other lepton components, there must be one quark component for every combination of flavour and colour, bringing the total to 24 (3 for charged leptons, 3 for neutrinos, and $2 \cdot 3 \cdot 3 = 18$ for quarks). Each of these is a four component bispinor, for a total of 96 complex-valued components for the fermion field.

An important definition is the barred fermion field $\bar{\psi}$ is defined to be $\psi^{\dagger}\gamma^{0}$, where $\dagger$ denotes the Hermitian adjoint and γ^{0} is the zeroth gamma matrix. If ψ is thought of as an $n \times 1$ matrix then $\bar{\psi}$ should be thought of as a $1 \times n$ matrix.

A Chiral Theory

An independent decomposition of ψ is that into chirality components:

"Left" chirality: $\psi = -(1-\gamma)\psi$

"Right" chirality: $\psi^{R} = \frac{1}{2}(1+\gamma_{5})\psi$

where γ_{5} is the fifth gamma matrix. This is very important in the Standard Model because *left and right chirality components are treated differently by the gauge interactions.*

In particular, under weak isospin SU(2) transformations the left-handed particles are weak-isospin doublets, whereas the right-handed are singlets – i.e. the weak isospin of ψ_{R} is zero. Put more simply, the weak interaction could rotate e.g. a left-handed electron into a left-handed neutrino (with emission of a W⁻), but could not do so with the same right-handed particles. As an aside, the right-handed neutrino originally did not exist in the standard model – but the discovery of neutrino oscillation implies that neutrinos must have mass, and since chirality can change during the propagation of a massive particle, right-handed neutrinos must exist in reality. This does not however change the (experimentally-proven) chiral nature of the weak interaction.

Furthermore, U(1) acts differently on ψ_{e}^{L} than on ψ_{e}^{R} (because they have different weak hypercharges).

Mass and Interaction Eigenstates

A distinction can thus be made between, for example, the mass and interaction eigenstates of the neutrino. The former is the state which propagates in free space, whereas the latter is the *different* state that participates in interactions. Which is the "fundamental" particle? For the neutrino, it is conventional to define the "flavour" (νe, νμ, or ντ) ( ) by the interaction eigenstate, whereas for the quarks we define the flavour (up, down, etc.) by the mass state. We can switch between these states using the CKM matrix for the quarks, or the PMNS matrix for the neutrinos (the charged leptons on the other hand are eigenstates of both mass and flavour).

As an aside, if a complex phase term exists within either of these matrices, it will give rise to direct CP violation, which could explain the dominance of matter over antimatter in our current universe. This has been proven for the CKM matrix, and is expected for the PMNS matrix.

Positive and Negative Energies

Finally, the quantum fields are sometimes decomposed into "positive" and "negative" energy parts: $\psi = \psi^+ + \psi^-$. This is not so common when a quantum field theory has been set up, but often features prominently in the process of quantizing a field theory.

Bosons

Due to the Higgs mechanism, the electroweak boson fields W_1, W_2, W_3, and B "mix" to create the states which are physically observable. To retain gauge invariance, the underlying fields must be massless, but the observable states can *gain masses* in the process. These states are:

The massive neutral (Z) boson:

$$Z = \cos\theta_W W_3 - \sin\theta_W B$$

The massless neutral boson:

$$A = \sin\theta_W W_3 + \cos\theta_W B$$

The massive charged W bosons:

$$W^\pm = \frac{1}{\sqrt{2}}\left(W_1 \mp iW_2\right)$$

where θ_W is the Weinberg angle.

The A field is the photon, which corresponds classically to the well-known electromagnetic four-potential – i.e. the electric and magnetic fields. The Z field actually contributes in every process the photon does, but due to its large mass, the contribution is usually negligible.

Perturbative QFT and the Interaction Picture

Much of the qualitative descriptions of the standard model in terms of "particles" and "forces" comes from the perturbative quantum field theory view of the model. In this, the Langrangian is decomposed as $\mathcal{L} = \mathcal{L}_0 + \mathcal{L}_1$ into separate *free field* and *interaction* Langrangians. The free fields care for particles in isolation, whereas processes involving several particles arise through interactions. The idea is that the state vector should only change when particles interact, meaning a free particle is one whose quantum state is constant. This corresponds to the interaction picture in quantum mechanics.

In the more common Schrödinger picture, even the states of free particles change over time: typically the phase changes at a rate which depends on their energy. In the alternative Heisenberg picture, state vectors are kept constant, at the price of having the operators (in particular the observables) be time-dependent. The interaction picture constitutes an intermediate between the two, where some time dependence is placed in the operators (the quantum fields) and some in the state vector. In QFT, the former is called the free field part of the model, and the latter is called the interaction part. The free field model can be solved exactly, and then the solutions to the full model can be expressed as perturbations of the free field solutions, for example using the Dyson series.

It should be observed that the decomposition into free fields and interactions is in principle arbitrary. For example, renormalization in QED modifies the mass of the free field electron to match that of a physical electron (with an electromagnetic field), and will in doing so add a term to the free field Lagrangian which must be cancelled by a counterterm in the interaction Lagrangian, that then shows up as a two-line vertex in the Feynman diagrams. This is also how the Higgs field is thought to give particles mass: the part of the interaction term which corresponds to the (nonzero) vacuum expectation value of the Higgs field is moved from the interaction to the free field Lagrangian, where it looks just like a mass term having nothing to do with Higgs.

Free Fields

Under the usual free/interaction decomposition, which is suitable for low energies, the free fields obey the following equations:

- The fermion field ψ satisfies the Dirac equation; $\partial - m_f c)\psi_f = 0$ for each type f of fermion.

- The photon field A satisfies the wave equation $\partial_\mu \partial^\mu A^\nu = 0$.

- The Higgs field φ satisfies the Klein–Gordon equation.

- The weak interaction fields Z, $W^\pm$ also satisfy the Proca equation.

These equations can be solved exactly. One usually does so by considering first solutions that are periodic with some period L along each spatial axis; later taking the limit: $L \to \infty$ will lift this periodicity restriction.

In the periodic case, the solution for a field F (any of the above) can be expressed as a Fourier series of the form

$$F(x) = \beta \sum_{\mathbf{p}} \sum_r E_{\mathbf{p}}^{-\frac{1}{2}} \left(a_r(\mathbf{p}) u_r(\mathbf{p}) e^{-\frac{ipx}{\hbar}} + b_r^\dagger(\mathbf{p}) v_r(\mathbf{p}) e^{\frac{ipx}{\hbar}} \right)$$

where:

- β is a normalization factor; for the fermion field ψ_f it is $\sqrt{m_f c^2 / V}$, , where $V = L^3$ is the volume of the fundamental cell considered; for the photon field A^μ it is $\hbar c / \sqrt{2V}$..

- The sum over p is over all momenta consistent with the period L, i.e., over all vectors $\dfrac{2\pi\hbar}{L}(n_1, n_2, n_3)$ where n_1, n_2, n_3 are integers.

- The sum over r covers other degrees of freedom specific for the field, such as polarization or spin; it usually comes out as a sum from 1 to 2 or from 1 to 3.

- E_p is the relativistic energy for a momentum p quantum of the field, $= \sqrt{m^2 c^4 + c^2 \mathbf{p}^2}$ when the rest mass is m.

- $a_r(\text{p})$ and $b_r^\dagger(\text{p})$ are annihilation and creation respectively operators for "a-particles" and "b-particles" respectively of momentum p; "b-particles" are the antiparticles of "a-particles". Different fields have different "a-" and "b-particles". For some fields, a and b are the same.

- $u_r(\text{p})$ and $v_r(\text{p})$ are non-operators which carry the vector or spinor aspects of the field (where relevant).

- $p = (E_\text{p} / c, \mathbf{p})$ is the four-momentum for a quantum with momentum p. $px = p_\mu x^\mu$ denotes an inner product of four-vectors.

In the limit $L \to \infty$, the sum would turn into an integral with help from the V hidden inside β. The numeric value of β also depends on the normalization chosen for $u_r(\mathbf{p})$ and $v_r(\mathbf{p})$.

Technically, $a_r^\dagger(\mathbf{p})$ is the Hermitian adjoint of the operator $a_r(\text{p})$ in the inner product space of ket vectors. The identification of $a_r^\dagger(\mathbf{p})$ and $a_r(\text{p})$ as creation and annihilation operators comes from comparing conserved quantities for a state before and after one of these have acted upon it. can for example be seen to add one particle, because it will add 1 to the eigenvalue of the a-particle number operator, and the momentum of that particle ought to be p since the eigenvalue of the vector-valued momentum operator increases by that much. For these derivations, one starts out with expressions for the operators in terms of the quantum fields. That the operators with † are creation operators and the one without annihilation operators is a convention, imposed by the sign of the commutation relations postulated for them.

An important step in preparation for calculating in perturbative quantum field theory is to separate the "operator" factors a and b above from their corresponding vector or spinor factors u and v. The vertices of Feynman graphs come from the way that u and v from different factors in the interaction Lagrangian fit together, whereas the edges come from the way that the as and bs must be moved around in order to put terms in the Dyson series on normal form.

Interaction terms and the path integral approach

The Lagrangian can also be derived without using creation and annihilation operators (the "canonical" formalism), by using a "path integral" approach, pioneered by Feynman building on the earlier work of Dirac. See e.g. A. Zee's QFT in a nutshell. This is one possible way that the Feynman diagrams, which are pictorial representations of interaction terms, can be derived relatively easily. A quick derivation is indeed presented at the article on Feynman diagrams.

Lagrangian Formalism

We can now give some more detail about the aforementioned free and interaction terms appearing in the Standard Model Lagrangian density. Any such term must be both gauge and reference-frame invariant, otherwise the laws of physics would depend on an arbitrary choice or the frame of an observer. Therefore, the global Poincaré symmetry, consisting of translational symmetry, rotational symmetry and the inertial reference frame invariance central to the theory of special relativity must apply. The local $SU(3) \times SU(2) \times U(1)$ gauge symmetry is the internal symmetry. The three factors of the gauge symmetry together give rise to the three fundamental interactions, after some appropriate relations have been defined, as we shall see.

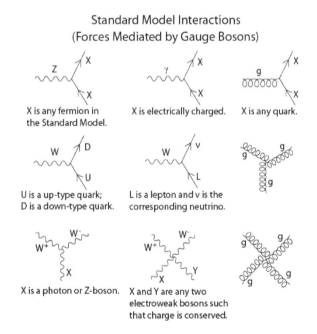

Standard Model Interactions
(Forces Mediated by Gauge Bosons)

X is any fermion in the Standard Model.

X is electrically charged.

X is any quark.

U is a up-type quark; D is a down-type quark.

L is a lepton and v is the corresponding neutrino.

X is a photon or Z-boson.

X and Y are any two electroweak bosons such that charge is conserved.

The above interactions show some basic interaction vertices – Feynman diagrams in the standard model are built from these vertices. Higgs boson interactions are however not shown, and neutrino oscillations are commonly added. The charge of the W bosons are dictated by the fermions they interact with.

A complete formulation of the Standard Model Lagrangian with all the terms written together can be found e.g. here.

Kinetic Terms

A free particle can be represented by a mass term, and a *kinetic* term which relates to the "motion" of the fields.

Fermion Fields

The kinetic term for a Dirac fermion is

$$i\bar{\psi}\gamma^\mu \partial_\mu \psi$$

where the notations are carried from earlier in the article. ψ can represent any, or all, Dirac fermions in the standard model. Generally, as below, this term is included within the couplings (creating an overall "dynamical" term).

Gauge Fields

For the spin-1 fields, first define the field strength tensor

$$F^a_{\mu\nu} = \partial_\mu A^a_\nu - \partial_\nu A^a_\mu + g f^{abc} A^b_\mu A^c_\nu$$

for a given gauge field (here we use A), with gauge coupling constant g. The quantity f^{abc} is the structure constant of the particular gauge group, defined by the commutator

$$[t_a, t_b] = i f^{abc} t_c,$$

where t_i are the generators of the group. In an Abelian (commutative) group (such as the U(1) we use here), since the generators t_a all commute with each other, the structure constants vanish. Of course, this is not the case in general – the standard model includes the non-Abelian SU(2) and SU(3) groups (such groups lead to what is called a Yang–Mills gauge theory).

We need to introduce three gauge fields corresponding to each of the subgroups SU(3) × SU(2) × U(1).

- The gluon field tensor will be denoted by $G^a_{\mu\nu}$, where the index a labels elements of the 8 representation of colour SU(3). The strong coupling constant is conventionally labelled g_s (or simply g where there is no ambiguity). *The observations leading to the discovery of this part of the Standard Model are discussed in the article in quantum chromodynamics.*

- The notation $W^a_{\mu\nu}$ will be used for the gauge field tensor of SU(2) where a runs over the 3 generators of this group. The coupling can be denoted g_w or again simply g. The gauge field will be denoted by W^a_μ.

- The gauge field tensor for the U(1) of weak hypercharge will be denoted by $B_{\mu\nu}$, the coupling by g', and the gauge field by B_μ.

The kinetic term can now be written simply as

$$\mathcal{L}_{\text{kin}} = -\frac{1}{4}B_{\mu\nu}B^{\mu\nu} - \frac{1}{2}\text{tr}W_{\mu\nu}W^{\mu\nu} - \frac{1}{2}\text{tr}G_{\mu\nu}G^{\mu\nu}$$

where the traces are over the SU(2) and SU(3) indices hidden in W and G respectively. The two-index objects are the field strengths derived from W and G the vector fields. There are also two extra hidden parameters: the theta angles for SU(2) and SU(3).

Coupling Terms

The next step is to "couple" the gauge fields to the fermions, allowing for interactions.

Electroweak Sector

The electroweak sector interacts with the symmetry group $U(1) \times SU(2)_L$, where the subscript L indicates coupling only to left-handed fermions.

$$\mathcal{L}_{\text{EW}} = \sum_{\psi}\bar{\psi}\gamma^{\mu}\left(i\partial_{\mu} - g'\frac{1}{2}Y_W B_{\mu} - g\frac{1}{2}\tau\mathbf{W}_{\mu}\right)\psi$$

Where B_{μ} is the U(1) gauge field; Y_W is the weak hypercharge (the generator of the U(1) group); W_{μ} is the three-component SU(2) gauge field; and the components of τ are the Pauli matrices (infinitesimal generators of the SU(2) group) whose eigenvalues give the weak isospin. Note that we have to redefine a new U(1) symmetry of *weak hypercharge*, different from QED, in order to achieve the unification with the weak force. The electric charge Q, third component of weak isospin T_3 (also called T_z, I_3 or I_z) and weak hypercharge Y_W are related by

$$Q = T_3 + \frac{1}{2}Y_W,$$

or by the alternate convention $Q = T_3 + Y_W$. The first convention (used in this article) is equivalent to the earlier Gell-Mann–Nishijima formula. We can then define the conserved current for weak isospin as

$$\mathbf{j}_{\mu} = \frac{1}{2}\bar{\psi}_L\gamma_{\mu}\tau\psi_L$$

and for weak hypercharge as

$$j_{\mu}^Y = 2(j_{\mu}^{em} - j_{\mu}^3)$$

where j_μ^{em} is the electric current and j_μ^3 the third weak isospin current. As explained above, *these currents mix* to create the physically observed bosons, which also leads to testable relations between the coupling constants.

To explain in a simpler way, we can see the effect of the electroweak interaction by picking out terms from the Lagrangian. We see that the SU(2) symmetry acts on each (left-handed) fermion doublet contained in ψ, for example

$$-\frac{g}{2}(\bar{v}_e\ \bar{e})\tau^+\gamma_\mu(W^-)^\mu\begin{pmatrix} v_e \\ e \end{pmatrix} = -\frac{g}{2}\bar{v}_e\gamma_\mu(W^-)^\mu e$$

where the particles are understood to be left-handed, and where

$$\tau^+ \equiv \frac{1}{2}(\tau^1 + i\tau^2) = \begin{pmatrix} 0 & 1 \\ 0 & 0 \end{pmatrix}$$

This is an interaction corresponding to a "rotation in weak isospin space" or in other words, a *transformation between e_L and v_{eL} via emission of a W$^-$ boson*. The U(1) symmetry, on the other hand, is similar to electromagnetism, but acts on all "*weak hypercharged*" fermions (both left and right handed) via the neutral Z^0, as well as the *charged* fermions via the photon.

Quantum Chromodynamics Sector

The quantum chromodynamics (QCD) sector defines the interactions between quarks and gluons, with SU(3) symmetry, generated by T_a. Since leptons do not interact with gluons, they are not affected by this sector. The Dirac Lagrangian of the quarks coupled to the gluon fields is given by

$$\mathcal{L}_{QCD} = i\bar{U}\left(\partial_\mu - ig_s G_\mu^a T^a\right)\gamma^\mu U + i\bar{D}\left(\partial_\mu - ig_s G_\mu^a T^a\right)\gamma^\mu D.$$

where D and U are the Dirac spinors associated with up- and down-type quarks, and other notations are continued from the previous section.

Mass Terms and the Higgs Mechanism

Mass Terms

The mass term arising from the Dirac Lagrangian (for any fermion ψ) is $-m\bar{\psi}\psi$ which is *not* invariant under the electroweak symmetry. This can be seen by writing ψ in terms of left and right handed components (skipping the actual calculation):

$$-m\bar{\psi}\psi = -m(\bar{\psi}_L\psi_R + \bar{\psi}_R\psi_L)$$

i.e. contribution from $\bar{\psi}_L \psi_L$ and $\bar{\psi}_R \psi_R$ terms do not appear. We see that the mass-generating interaction is achieved by constant flipping of particle chirality. The spin-half particles have no right/left chirality pair with the same SU(2) representations and equal and opposite weak hypercharges, so assuming these gauge charges are conserved in the vacuum, none of the spin-half particles could ever swap chirality, and must remain massless. Additionally, we know experimentally that the W and Z bosons are massive, but a boson mass term contains the combination e.g. $A^\mu A_\mu$, which clearly depends on the choice of gauge. Therefore, none of the standard model fermions *or* bosons can "begin" with mass, but must acquire it by some other mechanism.

The Higgs Mechanism

The solution to both these problems comes from the Higgs mechanism, which involves scalar fields (the number of which depend on the exact form of Higgs mechanism) which (to give the briefest possible description) are "absorbed" by the massive bosons as degrees of freedom, and which couple to the fermions via Yukawa coupling to create what looks like mass terms.

In the Standard Model, the Higgs field is a complex scalar of the group $SU(2)_L$:

$$\phi = \frac{1}{\sqrt{2}} \begin{pmatrix} \phi^+ \\ \phi^0 \end{pmatrix},$$

where the superscripts + and 0 indicate the electric charge (Q) of the components. The weak hypercharge (Y_W) of both components is 1.

The Higgs part of the Lagrangian is

$$\mathcal{L}_H = \left[\left(\partial_\mu - ig W_\mu^a t^a - ig' Y_\phi B_\mu \right) \phi \right]^2 + \mu^2 \phi^\dagger \phi - \lambda (\phi^\dagger \phi)^2,$$

where $\lambda > 0$ and $\mu^2 > 0$, so that the mechanism of spontaneous symmetry breaking can be used. There is a parameter here, at first hidden within the shape of the potential, that is very important. In a unitarity gauge one can set $\varphi^+ = 0$ and make φ^0 real. Then $\phi^0 = v$ is the non-vanishing vacuum expectation value of the Higgs field. v has units of mass, and it is the only parameter in the Standard Model which is not dimensionless. It is also much smaller than the Planck scale; it is approximately equal to the Higgs mass, and sets the scale for the mass of everything else. This is the only real fine-tuning to a small nonzero value in the Standard Model, and it is called the Hierarchy problem. Quadratic terms in W_μ and B_μ arise, which give masses to the W and Z bosons:

$$M_W = \frac{1}{2} v |g|$$

$$M_Z = \frac{1}{2} v \sqrt{g^2 + g'^2}$$

The Yukawa interaction terms are

$$\mathcal{L}_{YU} = \bar{U}_L G_u U_R \phi^0 - \bar{D}_L G_u U_R \phi^- + \bar{U}_L G_d D_R \phi^+ + \bar{D}_L G_d D_R \phi^0 + hc$$

where $G_{u,d}$ are 3×3 matrices of Yukawa couplings, with the ij term giving the coupling of the generations i and j.

Neutrino Masses

As previously mentioned, evidence shows neutrinos must have mass. But within the standard model, the right-handed neutrino does not exist, so even with a Yukawa coupling neutrinos remain massless. An obvious solution is to simply *add a right-handed neutrino* ν_R resulting in a Dirac mass term as usual. This field however must be a sterile neutrino, since being right-handed it experimentally belongs to an isospin singlet (T_3 = 0) and also has charge Q = 0, implying Y_W = 0 i.e. it does not even participate in the weak interaction. Current experimental status is that evidence for observation of sterile neutrinos is not convincing.

Another possibility to consider is that the neutrino satisfies the Majorana equation, which at first seems possible due to its zero electric charge. In this case the mass term is

$$-\frac{m}{2}\left(\bar{\nu}^C \nu + \bar{\nu}\nu^C\right)$$

where C denotes a charge conjugated (i.e. anti-) particle, and the terms are consistently all left (or all right) chirality (note that a left-chirality projection of an anti-particle is a right-handed field; care must be taken here due to different notations sometimes used). Here we are essentially flipping between LH neutrinos and RH anti-neutrinos (it is furthermore possible but *not* necessary that neutrinos are their own antiparticle, so these particles are the same). However, for the left-chirality neutrinos, this term changes weak hypercharge by 2 units - not possible with the standard Higgs interaction, requiring the Higgs field to be extended to include an extra triplet with weak hypercharge 2 - whereas for right-chirality neutrinos, no Higgs extensions are necessary. For both left and right chirality cases, Majorana terms violate lepton number, but possibly at a level beyond the current sensitivity of experiments to detect such violations.

It is possible to include both Dirac and Majorana mass terms in the same theory, which (in contrast to the Dirac-mass-only approach) can provide a "natural" explanation for the smallness of the observed neutrino masses, by linking the RH neutrinos to yet-unknown physics around the GUT scale.

Since in any case new fields must be postulated to explain the experimental results, neutrinos are an obvious gateway to searching physics beyond the Standard Model.

Detailed Information

This section provides more detail on some aspects, and some reference material.

Field Content in Detail

The Standard Model has the following fields. These describe one *generation* of leptons and quarks, and there are three generations, so there are three copies of each field. By CPT symmetry, there is a set of right-handed fermions with the opposite quantum numbers. The column "representation" indicates under which representations of the gauge groups that each field transforms, in the order (SU(3), SU(2), U(1)). Symbols used are common but not universal; superscript C denotes an antiparticle; and for the U(1) group, the value of the weak hypercharge is listed. Note that there are twice as many left-handed lepton field components as left-handed antilepton field components in each generation, but an equal number of left-handed quark and antiquark fields.

Free Parameters

Upon writing the most general Lagrangian without neutrinos, one finds that the dynamics depend on 19 parameters, whose numerical values are established by experiment. With neutrinos 7 more parameters are needed, 3 masses and 4 PMNS matrix parameters, for a total of 26 parameters. The neutrino parameter values are still uncertain. The 19 certain parameters are summarized here.

Parameters of the Standard Model			
Symbol	Description	Renormalization scheme (point)	Value
m_e	Electron mass		511 keV
m_μ	Muon mass		105.7 MeV
m_τ	Tau mass		1.78 GeV
m_u	Up quark mass	$\mu_{MS} = 2$ GeV	1.9 MeV
m_d	Down quark mass	$\mu_{MS} = 2$ GeV	4.4 MeV
m_s	Strange quark mass	$\mu_{MS} = 2$ GeV	87 MeV
m_c	Charm quark mass	$\mu_{MS} = m_c$	1.32 GeV
m_b	Bottom quark mass	$\mu_{MS} = m_b$	4.24 GeV
m_t	Top quark mass	On-shell scheme	172.7 GeV
θ_{12}	CKM 12-mixing angle		13.1°
θ_{23}	CKM 23-mixing angle		2.4°
θ_{13}	CKM 13-mixing angle		0.2°
δ	CKM CP-violating Phase		0.995
g_1 or g'	U(1) gauge coupling	$\mu_{MS} = m_Z$	0.357
g_2 or g	SU(2) gauge coupling	$\mu_{MS} = m_Z$	0.652
g_3 or g_s	SU(3) gauge coupling	$\mu_{MS} = m_Z$	1.221

θ_{QCD}	QCD vacuum angle		~0
v	Higgs vacuum expectation value		246 GeV
m_H	Higgs mass		~ 125 GeV (tentative)

The choice of free parameters is somewhat arbitrary. In the table above, gauge couplings are listed as free parameters, therefore with this choice Weinberg angle is not a free parameter - it is defined as $\tan\theta_W = \frac{g_1}{g_2}$. . Likewise, fine structure constant of QED is $\alpha = \frac{1}{4\pi}\frac{(g_1 g_2)^2}{g_1^2 + g_2^2}$..

Instead of fermion masses, dimensionless Yukawa couplings can be chosen as free parameters. For example, electron mass depends on the Yukawa coupling of electron to Higgs field, and its value is $m_e = \frac{y_e v}{\sqrt{2}}$..

Instead of the Higgs mass, the Higgs self-coupling strength $\lambda \sim 1/8$ can be chosen as a free parameter.

Additional Symmetries of the Standard Model

From the theoretical point of view, the Standard Model exhibits four additional global symmetries, not postulated at the outset of its construction, collectively denoted accidental symmetries, which are continuous U(1) global symmetries. The transformations leaving the Lagrangian invariant are:

$$\psi_q(x) \to e^{i\alpha/3}\psi_q$$

$$E_L \to e^{i\beta}E_L \text{ and } (e_R)^c \to e^{i\beta}(e_R)^c$$

$$M_L \to e^{i\beta}M_L \text{ and } (\mu_R)^c \to e^{i\beta}(\mu_R)^c$$

$$T_L \to e^{i\beta}T_L \text{ and } (\tau_R)^c \to e^{i\beta}(\tau_R)^c$$

The first transformation rule is shorthand meaning that all quark fields for all generations must be rotated by an identical phase simultaneously. The fields M_L, T_L and $(\mu_R)^c, (\tau_R)^c$ are the 2nd (muon) and 3rd (tau) generation analogs of E_L and $(e_R)^c$ fields.

By Noether's theorem, each symmetry above has an associated conservation law: the conservation of baryon number, electron number, muon number, and tau number. Each quark is assigned a baryon number of $\frac{1}{3}$, while each antiquark is assigned a baryon number of $-\frac{1}{3}$. Conservation of baryon number implies that the number of quarks minus the number of antiquarks is a constant. Within experimental limits, no violation of this conservation law has been found.

Similarly, each electron and its associated neutrino is assigned an electron number of +1, while the anti-electron and the associated anti-neutrino carry a −1 electron number.

Similarly, the muons and their neutrinos are assigned a muon number of +1 and the tau leptons are assigned a tau lepton number of +1. The Standard Model predicts that each of these three numbers should be conserved separately in a manner similar to the way baryon number is conserved. These numbers are collectively known as lepton family numbers (LF).

In addition to the accidental (but exact) symmetries described above, the Standard Model exhibits several approximate symmetries. These are the "SU(2) custodial symmetry" and the "SU(2) or SU(3) quark flavor symmetry."

Symmetries of the Standard Model and Associated Conservation Laws			
Symmetry	Lie Group	Symmetry Type	Conservation Law
Poincaré	Translations×SO(3,1)	Global symmetry	Energy, Momentum, Angular momentum
Gauge	SU(3)×SU(2)×U(1)	Local symmetry	Color charge, Weak isospin, Electric charge, Weak hypercharge
Baryon phase	U(1)	Accidental Global symmetry	Baryon number
Electron phase	U(1)	Accidental Global symmetry	Electron number
Muon phase	U(1)	Accidental Global symmetry	Muon number
Tau phase	U(1)	Accidental Global symmetry	Tau number

The U(1) Symmetry

For the leptons, the gauge group can be written $SU(2)_1 \times U(1)_L \times U(1)_R$. The two U(1) factors can be combined into $U(1)_Y \times U(1)_1$ where l is the lepton number. Gauging of the lepton number is ruled out by experiment, leaving only the possible gauge group $SU(2)_L \times U(1)_Y$. A similar argument in the quark sector also gives the same result for the electroweak theory.

The Charged and Neutral Current Couplings and Fermi Theory

The charged currents $j^{\pm} = j^1 \pm ij^2$ are

$$j_{\mu}^{+} = \bar{U}_{iL}\gamma_{\mu}D_{iL} + \bar{V}_{iL}\gamma_{\mu}l_{iL}.$$

These charged currents are precisely those that entered the Fermi theory of beta decay. The action contains the charge current piece

$$\mathcal{L}_{CC} = \frac{g}{\sqrt{2}}(j_{\mu}^{+}W^{-\mu} + j_{\mu}^{-}W^{+\mu}).$$

For energy much less than the mass of the W-boson, the effective theory becomes the current–current interaction of the Fermi theory.

However, gauge invariance now requires that the component W^3 of the gauge field also be coupled to a current that lies in the triplet of SU(2). However, this mixes with the U(1), and another current in that sector is needed. These currents must be uncharged in order to conserve charge. So we require the neutral currents

$$j_\mu^3 = \frac{1}{2}(\bar{U}_{iL}\gamma_\mu U_{iL} - \bar{D}_{iL}\gamma_\mu D_{iL} + \bar{\nu}_{iL}\gamma_\mu \nu_{iL} - \bar{l}_{iL}\gamma_\mu l_{iL})$$

$$j_\mu^{em} = \frac{2}{3}\bar{U}_i\gamma_\mu U_i - \frac{1}{3}\bar{D}_i\gamma_\mu D_i - \bar{l}_i\gamma_\mu l_i.$$

The neutral current piece in the Lagrangian is then

$$\mathcal{L}_{NC} = ej_\mu^{em}A^\mu + \frac{g}{\cos\theta_W}(J_\mu^3 - \sin^2\theta_W J_\mu^{em})Z^\mu.$$

Higgs Mechanism

In the Standard Model of particle physics, the Higgs mechanism is essential to explain the generation mechanism of the property "mass" for gauge bosons. Without the Higgs mechanism, or some other effect like it, all bosons (a type of fundamental particle) would be massless, but measurements show that the W+, W−, and Z bosons actually have relatively large masses of around 80 GeV/c². The Higgs field resolves this conundrum. The simplest description of the mechanism adds a quantum field (the Higgs field) that permeates all space, to the Standard Model. Below some extremely high temperature, the field causes spontaneous symmetry breaking during interactions. The breaking of symmetry triggers the Higgs mechanism, causing the bosons it interacts with to have mass. In the Standard Model, the phrase "Higgs mechanism" refers specifically to the generation of masses for the W±, and Z weak gauge bosons through electroweak symmetry breaking. The Large Hadron Collider at CERN announced results consistent with the Higgs particle on March 14, 2013, making it extremely likely that the field, or one like it, exists, and explaining how the Higgs mechanism takes place in nature.

The mechanism was proposed in 1962 by Philip Warren Anderson, following work in the late 1950s on symmetry breaking in superconductivity and a 1960 paper by Yoichiro Nambu that discussed its application within particle physics. A theory able to finally explain mass generation without "breaking" gauge theory was published almost simultaneously by three independent groups in 1964: by Robert Brout and François Englert; by Peter Higgs; and by Gerald Guralnik, C. R. Hagen, and Tom Kibble. The Higgs mechanism is therefore also called the Brout–Englert–Higgs mechanism or Englert–Brout–Higgs–Guralnik–Hagen–Kibble mechanism, Anderson–Higgs mechanism, Anderson–Higgs-Kibble mechanism, Higgs–Kibble mechanism by Abdus Salam and ABEGHHK'tH mechanism [for Anderson, Brout, Englert, Guralnik, Hagen, Higgs, Kibble and 't Hooft] by Peter Higgs.

On October 8, 2013, following the discovery at CERN's Large Hadron Collider of a new particle that appeared to be the long-sought Higgs boson predicted by the theory, it was announced that Peter Higgs and François Englert had been awarded the 2013 Nobel Prize in Physics (Englert's co-author Robert Brout had died in 2011 and the Nobel Prize is not usually awarded posthumously).

Standard Model

The Higgs mechanism was incorporated into modern particle physics by Steven Weinberg and Abdus Salam, and is an essential part of the standard model.

In the standard model, at temperatures high enough that electroweak symmetry is unbroken, all elementary particles are massless. At a critical temperature, the Higgs field becomes tachyonic; the symmetry is spontaneously broken by condensation, and the W and Z bosons acquire masses. (This also known as electroweak symmetry breaking; *EWSB*.)

Fermions, such as the leptons and quarks in the Standard Model, can also acquire mass as a result of their interaction with the Higgs field, but not in the same way as the gauge bosons.

Structure of the Higgs Field

In the standard model, the Higgs field is a SU(2) doublet (i.e. the standard representation with with two complex components called isospin), which is a scalar under Lorentz transformations. Its (weak hypercharge) U(1) charge is 1. Under U(1) rotations, it is multiplied by a phase, which thus mixes the real and imaginary parts of the complex spinor into each other—combining to the standard two component complex representation of the group U(2).

The Higgs field, through the interactions specified (summarized, represented, or even simulated) by its potential, induces spontaneous breaking of three out of the four generators ("directions") of the gauge group U(2). This is often written asSU(2) × U(1), (which is strictly speaking only the same on the level of infinitessimal symmetries) because the diagonal phase factor also acts on other fields in particular quarks. Three out of its four components would ordinarily amount to Goldstone bosons, if they were not coupled to gauge fields.

However, after symmetry breaking, these three of the four degrees of freedom in the Higgs field mix with the three W and Z bosons (W+, W− and Z), and are only observable as components of these weak bosons, which are now massive; while the one remaining degree of freedom becomes the Higgs boson—a new scalar particle.

The Photon as the Part that Remains Massless

The gauge group of the electroweak part of the standard model is SU(2) × U(1). The group SU(2) is the group of all 2-by-2 unitary matrices with unit determinant; all the orthonormal changes of coordinates in a complex two dimensional vector space.

Rotating the coordinates so that the second basis vector points in the direction of the Higgs boson makes the vacuum expectation value of H the spinor $(0, v)$. The generators for rotations about the x, y, and z axes are by half the Pauli matrices σ_x, σ_y, and σ_z, so that a rotation of angle θ about the z-axis takes the vacuum to

$$\left(0, ve^{-\frac{1}{2}i\theta}\right).$$

While the T_x and T_y generators mix up the top and bottom components of the spinor, the T_z rotations only multiply each by opposite phases. This phase can be undone by a U(1) rotation of angle $1/2\theta$. Consequently, under both an SU(2) T_z-rotation and a U(1) rotation by an amount $1/2\theta$, *the vacuum is invariant.*

This combination of generators

$$Q = T_z + \frac{Y}{2}$$

defines the unbroken part of the gauge group, where Q is the electric charge, T_z is the generator of rotations around the z-axis in the SU(2) and Y is the hypercharge generator of the U(1). This combination of generators (a z rotation in the SU(2) and a simultaneous U(1) rotation by half the angle) preserves the vacuum, and defines the unbroken gauge group in the standard model, namely *the electric charge* group. The part of the gauge field in this direction stays massless, and amounts to the physical photon.

Consequences for Fermions

In spite of the introduction of spontaneous symmetry breaking, the mass terms oppose the chiral gauge invariance. For these fields the mass terms should always be replaced by a gauge-invariant "Higgs" mechanism. One possibility is some kind of "Yukawa coupling" between the fermion field ψ and the Higgs field Φ, with unknown couplings G_ψ, which after symmetry breaking (more precisely: after expansion of the Lagrange density around a suitable ground state) again results in the original mass terms, which are now, however (i.e. by introduction of the Higgs field) written in a gauge-invariant way. The Lagrange density for the "Yukawa" interaction of a fermion field ψ and the Higgs field Φ is

$$\mathcal{L}_{\text{Fermion}}(\phi, A, \psi) = \bar{\psi}\gamma^\mu D_\mu \psi + G_\psi \bar{\psi}\phi\psi,$$

where again the gauge field A only enters D_μ (i.e., it is only indirectly visible). The quantities γ^μ are the Dirac matrices, and G_ψ is the already-mentioned "Yukawa" coupling parameter. Already now the mass-generation follows the same principle as above, namely from the existence of a finite expectation value $|\langle\phi\rangle|$, as described above. Again, this is crucial for the existence of the property "mass".

History of Research

Background

Spontaneous symmetry breaking offered a framework to introduce bosons into relativistic quantum field theories. However, according to Goldstone's theorem, these bosons should be massless. The only observed particles which could be approximately interpreted as Goldstone bosons were the pions, which Yoichiro Nambu related to chiral symmetry breaking.

A similar problem arises with Yang–Mills theory (also known as non-abelian gauge theory), which predicts massless spin-1 gauge bosons. Massless weakly interacting gauge bosons lead to long-range forces, which are only observed for electromagnetism and the corresponding massless photon. Gauge theories of the weak force needed a way to describe massive gauge bosons in order to be consistent.

Discovery

The mechanism was proposed in 1962 by Philip Warren Anderson, who discussed its consequences for particle physics but did not work out an explicit relativistic model. The relativistic model was developed in 1964 by three independent groups – Robert Brout and François Englert; Peter Higgs; and Gerald Guralnik, Carl Richard Hagen, and Tom Kibble. Slightly later, in 1965, but independently from the other publications the mechanism was also proposed by Alexander Migdal and Alexander Polyakov, at that time Soviet undergraduate students. However, the paper was delayed by the Editorial Office of JETP, and was published only in 1966.

Number six: Peter Higgs 2009

Philip W. Anderson, the first to propose the mechanism in 1962.

Five of the six 2010 APS Sakurai Prize Winners – (L to R) Tom Kibble, Gerald Guralnik, Carl Richard Hagen, François Englert, and Robert Brout

The mechanism is closely analogous to phenomena previously discovered by Yoichiro Nambu involving the "vacuum structure" of quantum fields in superconductivity. A similar but distinct effect (involving an affine realization of what is now recognized as the Higgs field), known as the Stueckelberg mechanism, had previously been studied by Ernst Stueckelberg.

These physicists discovered that when a gauge theory is combined with an additional field that spontaneously breaks the symmetry group, the gauge bosons can consistently acquire a nonzero mass. In spite of the large values involved this permits a gauge theory description of the weak force, which was independently developed by Steven Weinberg and Abdus Salam in 1967. Higgs's original article presenting the model was rejected by Physics Letters. When revising the article before resubmitting it to Physical Review Letters, he added a sentence at the end, mentioning that it implies the existence of one or more new, massive scalar bosons, which do not form complete representations of the symmetry group; these are the Higgs bosons.

The three papers by Brout and Englert; Higgs; and Guralnik, Hagen, and Kibble were each recognized as "milestone letters" by *Physical Review Letters* in 2008. While each of these seminal papers took similar approaches, the contributions and differences among the 1964 PRL symmetry breaking papers are noteworthy. All six physicists were jointly awarded the 2010 J. J. Sakurai Prize for Theoretical Particle Physics for this work.

Benjamin W. Lee is often credited with first naming the "Higgs-like" mechanism, although there is debate around when this first occurred. One of the first times the *Higgs* name appeared in print was in 1972 when Gerardus 't Hooft and Martinus J. G. Veltman referred to it as the "Higgs–Kibble mechanism" in their Nobel winning paper.

Examples

The Higgs mechanism occurs whenever a charged field has a vacuum expectation value. In the nonrelativistic context, this is the Landau model of a charged Bose–Einstein condensate, also known as a superconductor. In the relativistic condensate, the condensate is a scalar field, and is relativistically invariant.

Landau Model

The Higgs mechanism is a type of superconductivity which occurs in the vacuum. It occurs when all of space is filled with a sea of particles which are charged, or, in field language, when a charged field has a nonzero vacuum expectation value. Interaction with the quantum fluid filling the space prevents certain forces from propagating over long distances (as it does in a superconducting medium; e.g., in the Ginzburg–Landau theory).

A superconductor expels all magnetic fields from its interior, a phenomenon known as the Meissner effect. This was mysterious for a long time, because it implies that electromagnetic forces somehow become short-range inside the superconductor. Contrast this with the behavior of an ordinary metal. In a metal, the conductivity shields electric fields by rearranging charges on the surface until the total field cancels in the interior. But magnetic fields can penetrate to any distance, and if a magnetic monopole (an isolated magnetic pole) is surrounded by a metal the field can escape without collimating into a string. In a superconductor, however, electric charges move with no dissipation, and this allows for permanent surface currents, not just surface charges. When magnetic fields are introduced at the boundary of a superconductor, they produce surface currents which exactly neutralize them. The Meissner effect is due to currents in a thin surface layer, whose thickness, the London penetration depth, can be calculated from a simple model (the Ginzburg–Landau theory).

This simple model treats superconductivity as a charged Bose–Einstein condensate. Suppose that a superconductor contains bosons with charge q. The wavefunction of the bosons can be described by introducing a quantum field, ψ, which obeys the Schrödinger equation as a field equation (in units where the reduced Planck constant, $\hbar$, is set to 1):

$$i\frac{\partial}{\partial t}\psi = \frac{(\nabla - iqA)^2}{2m}\psi.$$

The operator $\psi(x)$ annihilates a boson at the point x, while its adjoint $\psi^\dagger$ creates a new boson at the same point. The wavefunction of the Bose–Einstein condensate is then the expectation value ψ of $\psi(x)$, which is a classical function that obeys the same equation. The interpretation of the expectation value is that it is the phase that one should give to a newly created boson so that it will coherently superpose with all the other bosons already in the condensate.

When there is a charged condensate, the electromagnetic interactions are screened. To see this, consider the effect of a gauge transformation on the field. A gauge transformation rotates the phase of the condensate by an amount which changes from point to point, and shifts the vector potential by a gradient:

$$\psi \rightarrow e^{iq\phi(x)}\psi$$
$$A \rightarrow A + \nabla\phi.$$

When there is no condensate, this transformation only changes the definition of the phase of ψ at every point. But when there is a condensate, the phase of the condensate defines a preferred choice of phase.

The condensate wave function can be written as

$$\psi(x) = \rho(x)e^{i\theta(x)},$$

where ρ is real amplitude, which determines the local density of the condensate. If the condensate were neutral, the flow would be along the gradients of θ, the direction in which the phase of the Schrödinger field changes. If the phase θ changes slowly, the flow is slow and has very little energy. But now θ can be made equal to zero just by making a gauge transformation to rotate the phase of the field.

The energy of slow changes of phase can be calculated from the Schrödinger kinetic energy,

$$H = \frac{1}{2m}|(qA + \nabla)\psi|^2,$$

and taking the density of the condensate ρ to be constant,

$$H \approx \frac{\rho^2}{2m}(qA + \nabla\theta)^2.$$

Fixing the choice of gauge so that the condensate has the same phase everywhere, the electromagnetic field energy has an extra term,

$$\frac{q^2 \rho^2}{2m} A^2.$$

When this term is present, electromagnetic interactions become short-ranged. Every field mode, no matter how long the wavelength, oscillates with a nonzero frequency. The lowest frequency can be read off from the energy of a long wavelength A mode,

$$E \approx \frac{\dot{A}^2}{2} + \frac{q^2 \rho^2}{2m} A^2.$$

This is a harmonic oscillator with frequency

$$\sqrt{\frac{1}{m} q^2 \rho^2}.$$

The quantity $|\psi|^2$ ($=\rho^2$) is the density of the condensate of superconducting particles.

In an actual superconductor, the charged particles are electrons, which are fermions not bosons. So in order to have superconductivity, the electrons need to somehow bind into Cooper pairs. The charge of the condensate q is therefore twice the electron charge e. The pairing in a normal superconductor is due to lattice vibrations, and is in fact very weak; this means that the pairs are very loosely bound. The description of a Bose–Einstein condensate of loosely bound pairs is actually more difficult than the description of a condensate of elementary particles, and was only worked out in 1957 by Bardeen, Cooper and Schrieffer in the famous BCS theory.

Abelian Higgs Mechanism

Gauge invariance means that certain transformations of the gauge field do not change the energy at all. If an arbitrary gradient is added to A, the energy of the field is exactly the same. This makes it difficult to add a mass term, because a mass term tends to push the field toward the value zero. But the zero value of the vector potential is not a gauge invariant idea. What is zero in one gauge is nonzero in another.

So in order to give mass to a gauge theory, the gauge invariance must be broken by a condensate. The condensate will then define a preferred phase, and the phase of the condensate will define the zero value of the field in a gauge-invariant way. The gauge-invariant definition is that a gauge field is zero when the phase change along any path from parallel transport is equal to the phase difference in the condensate wavefunction.

The condensate value is described by a quantum field with an expectation value, just as in the Ginzburg-Landau model.

In order for the phase of the vacuum to define a gauge, the field must have a phase (also referred to as 'to be charged'). In order for a scalar field Φ to have a phase, it must be complex, or (equivalently) it should contain two fields with a symmetry which rotates them into each other. The vector potential changes the phase of the quanta produced by the field when they move from point to point. In terms of fields, it defines how much to rotate the real and imaginary parts of the fields into each other when comparing field values at nearby points.

The only renormalizable model where a complex scalar field Φ acquires a nonzero value is the Mexican-hat model, where the field energy has a minimum away from zero. The action for this model is

$$S(\phi) = \int \frac{1}{2} |\partial \phi|^2 - \lambda \left(|\phi|^2 - \Phi^2 \right)^2,$$

which results in the Hamiltonian

$$H(\phi) = \frac{1}{2} |\dot{\phi}|^2 + |\nabla \phi|^2 + V(|\phi|).$$

The first term is the kinetic energy of the field. The second term is the extra potential energy when the field varies from point to point. The third term is the potential energy when the field has any given magnitude.

This potential energy, $V(z, \Phi) = \lambda(|z|^2 - \Phi^2)^2$, has a graph which looks like a Mexican hat, which gives the model its name. In particular, the minimum energy value is not at $z = 0$, but on the circle of points where the magnitude of z is Φ.

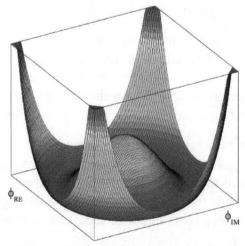

Higgs potential V. For a fixed value of λ the potential is presented upwards against the real and imaginary parts of Φ. The *Mexican-hat* or *champagne-bottle profile* at the ground should be noted.

When the field $\Phi(x)$ is not coupled to electromagnetism, the Mexican-hat potential has flat directions. Starting in any one of the circle of vacua and changing the phase of the field from point to point costs very little energy. Mathematically, if

$$\phi(x) = \Phi e^{i\theta(x)}$$

with a constant prefactor, then the action for the field $\theta(x)$, i.e., the "phase" of the Higgs field $\Phi(x)$, has only derivative terms. This is not a surprise. Adding a constant to $\theta(x)$ is a symmetry of the original theory, so different values of $\theta(x)$ cannot have different energies. This is an example of Goldstone's theorem: spontaneously broken continuous symmetries normally produce massless excitations.

The Abelian Higgs model is the Mexican-hat model coupled to electromagnetism:

$$S(\phi, A) = \int -\frac{1}{4} F^{\mu\nu} F_{\mu\nu} + |(\partial - iqA)\phi|^2 - \lambda(|\phi|^2 - \Phi^2)^2.$$

The classical vacuum is again at the minimum of the potential, where the magnitude of the complex field φ is equal to Φ. But now the phase of the field is arbitrary, because gauge transformations change it. This means that the field $\theta(x)$ can be set to zero by a gauge transformation, and does not represent any actual degrees of freedom at all.

Furthermore, choosing a gauge where the phase of the vacuum is fixed, the potential energy for fluctuations of the vector field is nonzero. So in the abelian Higgs model, the gauge field acquires a mass. To calculate the magnitude of the mass, consider a constant value of the vector potential A in the x direction in the gauge where the condensate has constant phase. This is the same as a sinusoidally varying condensate in the gauge where the vector potential is zero. In the gauge where A is zero, the potential energy density in the condensate is the scalar gradient energy:

$$E = \frac{1}{2}\left|\partial\left(\Phi e^{iqAx}\right)\right|^2 = \frac{1}{2} q^2 \Phi^2 A^2.$$

This energy is the same as a mass term $1/2m^2A^2$ where $m = q\Phi$.

Nonabelian Higgs Mechanism

The Nonabelian Higgs model has the following action:

$$S(\phi, \mathbf{A}) = \int \frac{1}{4g^2} \text{tr}(F^{\mu\nu} F_{\mu\nu}) + |D\phi|^2 + V(|\phi|)$$

where now the nonabelian field A is contained in the covariant derivative D and in the tensor components $F^{\mu\nu}$ and $F_{\mu\nu}$ (the relation between A and those components is well-known from the Yang–Mills theory).

It is exactly analogous to the Abelian Higgs model. Now the field φ is in a representation of the gauge group, and the gauge covariant derivative is defined by the rate of change of the field minus the rate of change from parallel transport using the gauge field A as a connection.

$$D\phi = \partial\phi - iA^k t_k \phi$$

Again, the expectation value of Φ defines a preferred gauge where the vacuum is constant, and fixing this gauge, fluctuations in the gauge field A come with a nonzero energy cost.

Depending on the representation of the scalar field, not every gauge field acquires a mass. A simple example is in the renormalizable version of an early electroweak model due to Julian Schwinger. In this model, the gauge group is SO(3) (or SU(2) – there are no spinor representations in the model), and the gauge invariance is broken down to U(1) or SO(2) at long distances. To make a consistent renormalizable version using the Higgs mechanism, introduce a scalar field φ^a which transforms as a vector (a triplet) of SO(3). If this field has a vacuum expectation value, it points in some direction in field space. Without loss of generality, one can choose the z-axis in field space to be the direction that φ is pointing, and then the vacuum expectation value of φ is (0, 0, A), where A is a constant with dimensions of mass ($c = \hbar = 1$).

Rotations around the z-axis form a U(1) subgroup of SO(3) which preserves the vacuum expectation value of φ, and this is the unbroken gauge group. Rotations around the x and y-axis do not preserve the vacuum, and the components of the SO(3) gauge field which generate these rotations become massive vector mesons. There are two massive W mesons in the Schwinger model, with a mass set by the mass scale A, and one massless U(1) gauge boson, similar to the photon.

The Schwinger model predicts magnetic monopoles at the electroweak unification scale, and does not predict the Z meson. It doesn't break electroweak symmetry properly as in nature. But historically, a model similar to this (but not using the Higgs mechanism) was the first in which the weak force and the electromagnetic force were unified.

Affine Higgs Mechanism

Ernst Stueckelberg discovered a version of the Higgs mechanism by analyzing the theory of quantum electrodynamics with a massive photon. Effectively, Stueckelberg's model is a limit of the regular Mexican hat Abelian Higgs model, where the vacuum expectation value H goes to infinity and the charge of the Higgs field goes to zero in such a way that their product stays fixed. The mass of the Higgs boson is proportional to H, so the Higgs boson becomes infinitely massive and decouples, so is not present in the discussion. The vector meson mass, however, equals to the product eH, and stays finite.

The interpretation is that when a U(1) gauge field does not require quantized charges, it is possible to keep only the angular part of the Higgs oscillations, and discard the radial part. The angular part of the Higgs field θ has the following gauge transformation law:

$$\theta \;\; \rightarrow \theta + e\alpha$$
$$A \;\; \rightarrow A + \partial\alpha.$$

The gauge covariant derivative for the angle (which is actually gauge invariant) is:

$$D\theta = \partial\theta - eAH.$$

In order to keep θ fluctuations finite and nonzero in this limit, θ should be rescaled by H, so that its kinetic term in the action stays normalized. The action for the theta field is read off from the Mexican hat action by substituting $\phi = He^{\frac{1}{H}i\theta}$.

$$S = \int \frac{1}{4}F^2 + \frac{1}{2}(D\theta)^2 = \int \frac{1}{4}F^2 + \frac{1}{2}(\partial\theta - HeA)^2 = \int \frac{1}{4}F^2 + \frac{1}{2}(\partial\theta - mA)^2 .$$

since eH is the gauge boson mass. By making a gauge transformation to set $\theta = 0$, the gauge freedom in the action is eliminated, and the action becomes that of a massive vector field:

$$S = \int \frac{1}{4}F^2 + \frac{1}{2}m^2A^2 .$$

To have arbitrarily small charges requires that the U(1) is not the circle of unit complex numbers under multiplication, but the real numbers R under addition, which is only different in the global topology. Such a U(1) group is *non-compact*. The field θ transforms as an affine representation of the gauge group. Among the allowed gauge groups, only non-compact U(1) admits affine representations, and the U(1) of electromagnetism is experimentally known to be compact, since charge quantization holds to extremely high accuracy.

The Higgs condensate in this model has infinitesimal charge, so interactions with the Higgs boson do not violate charge conservation. The theory of quantum electrodynamics with a massive photon is still a renormalizable theory, one in which electric charge is still conserved, but magnetic monopoles are not allowed. For nonabelian gauge theory, there is no affine limit, and the Higgs oscillations cannot be too much more massive than the vectors.

Cabibbo–Kobayashi–Maskawa Matrix

In the Standard Model of particle physics, the Cabibbo–Kobayashi–Maskawa matrix (CKM matrix, quark mixing matrix, sometimes also called KM matrix) is a unitary matrix which contains information on the strength of flavour-changing weak decays. Technically, it specifies the mismatch of quantum states of quarks when they propagate freely and when they take part in the weak interactions. It is important in the understanding of CP violation. This matrix was introduced for three generations of quarks by Makoto Kobayashi and Toshihide Maskawa, adding one generation to the matrix previously introduced by Nicola Cabibbo. This matrix is also an extension of the GIM mechanism, which only includes two of the three current families of quarks.

The Matrix

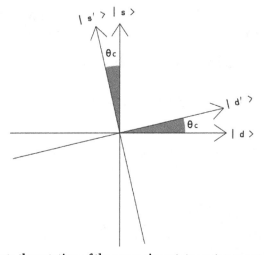

The Cabibbo angle represents the rotation of the mass eigenstate vector space formed by the mass eigenstates into the weak eigenstate vector space formed by the weak eigenstates . $\theta_c = 13.02°$.

In 1963, Nicola Cabibbo introduced the Cabibbo angle (θ_c) to preserve the universality of the weak interaction. Cabibbo was inspired by previous work by Murray Gell-Mann and Maurice Lévy, on the effectively rotated nonstrange and strange vector and axial weak currents, which he references.

In light of current knowledge (quarks were not yet theorized), the Cabibbo angle is related to the relative probability that down and strange quarks decay into up quarks ($|V_{ud}|^2$ and $|V_{us}|^2$ respectively). In particle physics parlance, the object that couples to the up quark via charged-current weak interaction is a superposition of down-type quarks, here denoted by d'. Mathematically this is:

$$d' = V_{ud}d + V_{us}s,$$

or using the Cabibbo angle:

$$d' = \cos\theta_c d + \sin\theta_c s.$$

Using the currently accepted values for $|V_{ud}|$ and $|V_{us}|$, the Cabibbo angle can be calculated using

$$\tan\theta_c = \frac{|V_{us}|}{|V_{ud}|} = \frac{0.22534}{0.97427} \rightarrow \theta_c = 13.02°.$$

When the charm quark was discovered in 1974, it was noticed that the down and strange quark could decay into either the up or charm quark, leading to two sets of equations:

$$d' = V_{ud}d + V_{us}s;$$

$$s' \quad V_{cd}d \quad V_{cs}s$$

or using the Cabibbo angle:

$$d' = \cos\theta_c d + \sin\theta_c s;$$

$$s' = -\sin\theta_c d + \cos\theta_c s.$$

This can also be written in matrix notation as:

$$\begin{bmatrix} d' \\ s' \end{bmatrix} = \begin{bmatrix} V_{ud} & V_{us} \\ V_{cd} & V_{cs} \end{bmatrix} \begin{bmatrix} d \\ s \end{bmatrix},$$

or using the Cabibbo angle

$$\begin{bmatrix} d' \\ s' \end{bmatrix} = \begin{bmatrix} \cos\theta_c & \sin\theta_c \\ -\sin\theta_c & \cos\theta_c \end{bmatrix} \begin{bmatrix} d \\ s \end{bmatrix},$$

where the various $|V_{ij}|^2$ represent the probability that the quark of j flavor decays into a quark of i flavor. This 2×2 rotation matrix is called the Cabibbo matrix.

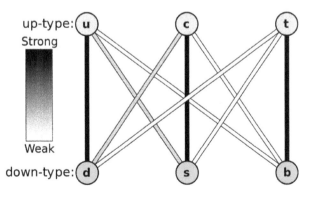

A pictorial representation of the six quarks' decay modes, with mass increasing from left to right.

Observing that CP-violation could not be explained in a four-quark model, Kobayashi and Maskawa generalized the Cabibbo matrix into the Cabibbo–Kobayashi–Maskawa matrix (or CKM matrix) to keep track of the weak decays of three generations of quarks:

$$\begin{bmatrix} d' \\ s' \\ b' \end{bmatrix} = \begin{bmatrix} V_{ud} & V_{us} & V_{ub} \\ V_{cd} & V_{cs} & V_{cb} \\ V_{td} & V_{ts} & V_{tb} \end{bmatrix} \begin{bmatrix} d \\ s \\ b \end{bmatrix}.$$

On the left is the weak interaction doublet partners of up-type quarks, and on the right is the CKM matrix along with a vector of mass eigenstates of down-type quarks. The CKM matrix describes the probability of a transition from one quark i to another quark j. These transitions are proportional to $|V_{ij}|^2$.

Currently, the best determination of the magnitudes of the CKM matrix elements is:

$$\begin{bmatrix} |V_{ud}| & |V_{us}| & |V_{ub}| \\ |V_{cd}| & |V_{cs}| & |V_{cb}| \\ |V_{td}| & |V_{ts}| & |V_{tb}| \end{bmatrix} = \begin{bmatrix} 0.97427 \pm 0.00015 & 0.22534 \pm 0.00065 & 0.00351^{+0.00015}_{-0.00014} \\ 0.22520 \pm 0.00065 & 0.97344 \pm 0.00016 & 0.0412^{+0.0011}_{-0.0005} \\ 0.00867^{+0.00029}_{-0.00031} & 0.0404^{+0.0011}_{-0.0005} & 0.999146^{+0.000021}_{-0.000046} \end{bmatrix}.$$

Note that the choice of usage of down-type quarks in the definition is purely arbitrary and does not represent some sort of deep physical asymmetry between up-type and down-type quarks. We could just as easily define the matrix the other way around, describing weak interaction partners of mass eigenstates of up-type quarks, u', c' and t', in terms of u, c, and t. Since the CKM matrix is unitary (and therefore its inverse is the same as its conjugate transpose), we would obtain essentially the same matrix.

Counting

To proceed further, it is necessary to count the number of parameters in this matrix, V which appear in experiments, and therefore are physically important. If there are N generations of quarks ($2N$ flavours) then

- An $N \times N$ unitary matrix (that is, a matrix V such that $VV^\dagger = I$, where $V^\dagger$ is the conjugate transpose of V and I is the identity matrix) requires N^2 real parameters to be specified.

- $2N - 1$ of these parameters are not physically significant, because one phase can be absorbed into each quark field (both of the mass eigenstates, and of the weak eigenstates), but an overall common phase is unobservable. Hence, the total number of free variables independent of the choice of the phases of basis vectors is $N^2 - (2N - 1) = (N - 1)^2$.

  o Of these, $N(N - 1)/2$ are rotation angles called quark *mixing angles*.

  o The remaining $(N - 1)(N - 2)/2$ are complex phases, which cause CP violation.

For the case $N = 2$, there is only one parameter which is a mixing angle between two generations of quarks. Historically, this was the first version of CKM matrix when only two generations were known. It is called the Cabibbo angle after its inventor Nicola Cabibbo.

For the Standard Model case ($N = 3$), there are three mixing angles and one CP-violating complex phase.

Observations and Predictions

Cabibbo's idea originated from a need to explain two observed phenomena:

1. the transitions $u \leftrightarrow d$, $e \leftrightarrow \nu_e$, and $\mu \leftrightarrow \nu_\mu$ had similar amplitudes.

2. the transitions with change in strangeness $\Delta S = 1$ had amplitudes equal to $1/4$ of those with $\Delta S = 0$.

Cabibbo's solution consisted of postulating weak universality to resolve the first issue, along with a mixing angle θ_c, now called the *Cabibbo angle*, between the d and s quarks to resolve the second.

For two generations of quarks, there are no CP violating phases, as shown by the counting of the previous section. Since CP violations were seen in neutral kaon decays already in 1964, the emergence of the Standard Model soon after was a clear signal of the existence of a third generation of quarks, as pointed out in 1973 by Kobayashi and Maskawa. The discovery of the bottom quark at Fermilab (by Leon Lederman's group) in 1976 therefore immediately started off the search for the missing third-generation quark, the top quark.

Note, however, that the specific values of the angles are *not* a prediction of the standard model: they are open, unfixed parameters. At this time, there is no generally accepted theory that explains why the measured values are what they are.

Weak Universality

The constraints of unitarity of the CKM-matrix on the diagonal terms can be written as

$$\sum_k |V_{ik}|^2 = \sum_i |V_{ik}|^2 = 1$$

for all generations i. This implies that the sum of all couplings of any of the up-type quarks to all the down-type quarks is the same for all generations. This relation is called *weak universality* and was first pointed out by Nicola Cabibbo in 1967. Theoretically it is a consequence of the fact that all SU(2) doublets couple with the same strength to the vector bosons of weak interactions. It has been subjected to continuing experimental tests.

The Unitarity Triangles

The remaining constraints of unitarity of the CKM-matrix can be written in the form

$$\sum_k V_{ik} V_{jk}^* = 0.$$

For any fixed and different i and j, this is a constraint on three complex numbers, one for each k, which says that these numbers form the sides of a triangle in the complex plane. There are six choices of i and j (three independent), and hence six such triangles, each of which is called a *unitary triangle*. Their shapes can be very different, but they all have the same area, which can be related to the CP violating phase. The area vanishes for the specific parameters in the Standard Model for which there would be no CP violation. The orientation of the triangles depend on the phases of the quark fields.

Since the three sides of the triangles are open to direct experiment, as are the three angles, a class of tests of the Standard Model is to check that the triangle closes. This is the purpose of a modern series of experiments under way at the Japanese BELLE and the American BaBar experiments, as well as at LHCb in CERN, Switzerland.

Parameterizations

Four independent parameters are required to fully define the CKM matrix. Many parameterizations have been proposed, and three of the most common ones are shown below.

KM Parameters

The original parameterization of Kobayashi and Maskawa used three angles ($\theta_1, \theta_2, \theta_3$) and a CP-violating phase (δ). Cosines and sines of the angles are denoted c_i and s_i, respectively. θ_1 is the Cabibbo angle.

$$\begin{bmatrix} c_1 & -s_1 c_3 & -s_1 s_3 \\ s_1 c_2 & c_1 c_2 c_3 - s_2 s_3 e^{i\delta} & c_1 c_2 s_3 + s_2 c_3 e^{i\delta} \\ s_1 s_2 & c_1 s_2 c_3 + c_2 s_3 e^{i\delta} & c_1 s_2 s_3 - c_2 c_3 e^{i\delta} \end{bmatrix}.$$

"Standard" Parameters

A "standard" parameterization of the CKM matrix uses three Euler angles ($\theta_{12}, \theta_{23}, \theta_{13}$) and one CP-violating phase (δ_{13}). Couplings between quark generation i and j vanish if $\theta_{ij} = 0$. Cosines and sines of the angles are denoted c_{ij} and s_{ij}, respectively. θ_{12} is the Cabibbo angle.

$$\begin{bmatrix} 1 & 0 & 0 \\ 0 & c_{23} & s_{23} \\ 0 & -s_{23} & c_{23} \end{bmatrix} \begin{bmatrix} c_{13} & 0 & s_{13}e^{-i\delta_{13}} \\ 0 & 1 & 0 \\ -s_{13}e^{i\delta_{13}} & 0 & c_{13} \end{bmatrix} \begin{bmatrix} c_{12} & s_{12} & 0 \\ -s_{12} & c_{12} & 0 \\ 0 & 0 & 1 \end{bmatrix}$$

$$= \begin{bmatrix} c_{12}c_{13} & s_{12}c_{13} & s_{13}e^{-i\delta_{13}} \\ -s_{12}c_{23} - c_{12}s_{23}s_{13}e^{i\delta_{13}} & c_{12}c_{23} - s_{12}s_{23}s_{13}e^{i\delta_{13}} & s_{23}c_{13} \\ s_{12}s_{23} - c_{12}c_{23}s_{13}e^{i\delta_{13}} & -c_{12}s_{23} - s_{12}c_{23}s_{13}e^{i\delta_{13}} & c_{23}c_{13} \end{bmatrix}.$$

The currently best known values for the standard parameters are:

θ_{12} = 13.04±0.05°, θ_{13} = 0.201±0.011°, θ_{23} = 2.38±0.06°, and δ_{13} = 1.20±0.08 rad.

Wolfenstein Parameters

A third parameterization of the CKM matrix was introduced by Lincoln Wolfenstein with the four parameters λ, A, ρ, and η. The four Wolfenstein parameters have the property that all are of order 1 and are related to the "standard" parameterization:

$\lambda = s_{12}$

$A\lambda^2 = s_{23}$

$A\lambda^3(\rho - i\eta) = s_{13}e^{-i\delta}$

The Wolfenstein parameterization of the CKM matrix, is an approximation of the standard parameterization. To order λ^3, it is:

$$\begin{bmatrix} 1 - \lambda^2/2 & \lambda & A\lambda^3(\rho - i\eta) \\ -\lambda & 1 - \lambda^2/2 & A\lambda^2 \\ A\lambda^3(1 - \rho - i\eta) & -A\lambda^2 & 1 \end{bmatrix}.$$

The CP violation can be determined by measuring $\rho - i\eta$.

Using the values of the previous section for the CKM matrix, the best determination of the Wolfenstein parameters is:

$\lambda = 0.2257^{+0.0009}_{-0.0010}$,

$A = 0.814^{+0.021}_{-0.022}$,

$\rho = 0.135^{+0.031}_{-0.016}$,

and $\eta = 0.349^{+0.015}_{-0.017}$.

Nobel Prize

In 2008, Kobayashi and Maskawa shared one half of the Nobel Prize in Physics "for the discovery of the origin of the broken symmetry which predicts the existence of at least three families of quarks in nature". Some physicists were reported to harbor bitter feelings about the fact that the Nobel Prize committee failed to reward the work of Cabibbo, whose prior work was closely related to that of Kobayashi and Maskawa. Asked for a reaction on the prize, Cabibbo preferred to give no comment.

Spontaneous Symmetry Breaking

Spontaneous symmetry breaking is a mode of realization of symmetry breaking in a physical system, where the underlying laws are invariant under a symmetry transformation, but the system as a whole changes under such transformations, in contrast to explicit symmetry breaking. It is a spontaneous process by which a system in a symmetrical state ends up in an asymmetrical state. It thus describes systems where the equations of motion or the Lagrangian obey certain symmetries, but the lowest-energy solutions do not exhibit that symmetry.

Consider a symmetrical upward dome with a trough circling the bottom. If a ball is put at the very peak of the dome, the system is symmetrical with respect to a rotation around the center axis. But the ball may *spontaneously break* this symmetry by rolling down the dome into the trough, a point of lowest energy. Afterward, the ball has come to a rest at some fixed point on the perimeter. The dome and the ball retain their individual symmetry, but the system does not.

Most simple phases of matter and phase transitions, like crystals, magnets, and conventional superconductors can be simply understood from the viewpoint of spontaneous symmetry breaking. Notable exceptions include topological phases of matter like the fractional quantum Hall effect.

Spontaneous Symmetry Breaking in Physics

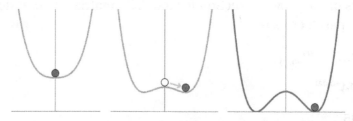

Spontaneous symmetry breaking illustrated: At high energy levels (*left*) the ball settles in the center, and the result is symmetrical. At lower energy levels (*right*), the overall "rules" remain symmetrical, but the symmetric "Mexican hat" enforces an asymmetric outcome, since eventually the ball must rest at some random spot on the bottom, "spontaneously", and not all others.

Particle Physics

In particle physics the force carrier particles are normally specified by field equations with gauge symmetry; their equations predict that certain measurements will be the same at any point in the field. For instance, field equations might predict that the mass of two quarks is constant. Solving the equations to find the mass of each quark might give two solutions. In one solution, quark A is heavier than quark B. In the second solution, quark B is heavier than quark A *by the same amount*. The symmetry of the equations is not reflected by the individual solutions, but it is reflected by the range of solutions.

An actual measurement reflects only one solution, representing a breakdown in the symmetry of the underlying theory. "Hidden" is a better term than "broken", because the symmetry is always there in these equations. This phenomenon is called *spontaneous* symmetry breaking (SSB) because *nothing* (that we know of) breaks the symmetry in the equations.

Chiral Symmetry

Chiral symmetry breaking is an example of spontaneous symmetry breaking affecting the chiral symmetry of the strong interactions in particle physics. It is a property of quantum chromodynamics, the quantum field theory describing these interactions, and is responsible for the bulk of the mass (over 99%) of the nucleons, and thus of all common matter, as it converts very light bound quarks into 100 times heavier constituents of baryons. The approximate Nambu–Goldstone bosons in this spontaneous symmetry breaking process are the pions, whose mass is an order of magnitude lighter than the mass of the nucleons. It served as the prototype and significant ingredient of the Higgs mechanism underlying the electroweak symmetry breaking.

Higgs Mechanism

The strong, weak, and electromagnetic forces can all be understood as arising from gauge symmetries. The Higgs mechanism, the spontaneous symmetry breaking of gauge symmetries, is an important component in understanding the superconductivity of metals and the origin of particle masses in the standard model of particle physics. One important consequence of the distinction between true symmetries and *gauge symmetries*, is that the spontaneous breaking of a gauge symmetry does not give rise to characteristic massless Nambu–Goldstone physical modes, but only massive modes, like the plasma mode in a superconductor, or the Higgs mode observed in particle physics.

In the standard model of particle physics, spontaneous symmetry breaking of the SU(2) × U(1) gauge symmetry associated with the electro-weak force generates masses for several particles, and separates the electromagnetic and weak forces. The W and Z bosons are the elementary particles that mediate the weak interaction, while the photon mediates the electromagnetic interaction. At energies much greater than 100 GeV all these particles behave in a similar manner. The Weinberg–Salam theory predicts that, at lower energies, this symmetry is broken so that the photon and the massive W and Z bosons emerge. In addition, fermions develop mass consistently.

Without spontaneous symmetry breaking, the Standard Model of elementary particle interactions requires the existence of a number of particles. However, some particles (the W and Z bosons) would then be predicted to be massless, when, in reality, they are observed to have mass. To overcome this, spontaneous symmetry breaking is augmented by the Higgs mechanism to give these particles mass. It also suggests the presence of a new particle, the Higgs boson, detected in 2012.

Superconductivity of metals is a condensed-matter analog of the Higgs phenomena, in which a condensate of Cooper pairs of electrons spontaneously breaks the U(1) gauge symmetry associated with light and electromagnetism.

Condensed Matter Physics

Most phases of matter can be understood through the lens of spontaneous symmetry breaking. For example, crystals are periodic arrays of atoms that are not invariant under all translations (only under a small subset of translations by a lattice vector). Magnets have north and south poles that are oriented in a specific direction, breaking rotational symmetry. In addition to these examples, there are a whole host of other symmetry-breaking phases of matter including nematic phases of liquid crystals, charge- and spin-density waves, superfluids and many others.

There are several known examples of matter that cannot be described by spontaneous symmetry breaking, including: topologically ordered phases of matter like fractional quantum Hall liquids, and spin-liquids. These states do not break any symmetry, but are distinct phases of matter. Unlike the case of spontaneous symmetry breaking, there is not a general framework for describing such states.

Continuous Symmetry

The ferromagnet is the canonical system which spontaneously breaks the continuous symmetry of the spins below the Curie temperature and at $h = 0$, where h is the external magnetic field. Below the Curie temperature the energy of the system is invariant under inversion of the magnetization $m(x)$ such that $m(x) = -m(-x)$. The symmetry is spontaneously broken as $h \to 0$ when the Hamiltonian becomes invariant under the inversion transformation, but the expectation value is not invariant.

Spontaneously, symmetry broken phases of matter are characterized by an order parameter that describes the quantity which breaks the symmetry under consideration. For example, in a magnet, the order parameter is the local magnetization.

Spontaneously breaking of a continuous symmetry is inevitably accompanied by gapless (meaning that these modes do not cost any energy to excite) Nambu–Goldstone modes associated with slow long-wavelength fluctuations of the order parameter. For example, vibrational modes in a crystal, known as phonons, are associated with slow density fluctuations of the crystal's atoms. The associated Goldstone mode for magnets are oscillating waves of spin known as spin-waves. For symmetry-breaking states, whose order parameter is not a conserved quantity, Nambu–Goldstone modes are typically massless and propagate at a constant velocity.

An important theorem, due to Mermin and Wagner, states that, at finite temperature, thermally activated fluctuations of Nambu–Goldstone modes destroy the long-range order, and prevent spontaneous symmetry breaking in one- and two-dimensional systems. Similarly, quantum fluctuations of the order parameter prevent most types of continuous symmetry breaking in one-dimensional systems even at zero temperature (an important exception is ferromagnets, whose order parameter, magnetization, is an exactly conserved quantity and does not have any quantum fluctuations).

Other long-range interacting systems such as cylindrical curved surfaces interacting via the Coulomb potential or Yukawa potential has been shown to break translational and rotational symmetries. It was shown, in the presence of a symmetric Hamiltonian, and in the limit of infinite volume, the system spontaneously adopts a chiral configuration, i.e. breaks mirror plane symmetry.

Dynamical Symmetry Breaking

Dynamical symmetry breaking (DSB) is a special form of spontaneous symmetry breaking where the ground state of the system has reduced symmetry properties compared to its theoretical description (Lagrangian).

Dynamical breaking of a global symmetry is a spontaneous symmetry breaking, that happens not at the (classical) tree level (i.e. at the level of the bare action), but due to quantum corrections (i.e. at the level of the effective action).

Dynamical breaking of a gauge symmetry is subtler. In the conventional spontaneous gauge symmetry breaking, there exists an unstable Higgs particle in the theory, which drives the vacuum to a symmetry-broken phase (see e.g. Electroweak interaction). In dynamical gauge symmetry breaking, however, no unstable Higgs particle operates in the theory, but the bound states of the system itself provide the unstable fields that render the phase transition. For example, Bardeen, Hill, and Lindner published a paper which attempts to replace the conventional Higgs mechanism in the standard model, by a DSB that is driven by a bound state of top-antitop quarks (such models, where a composite particle plays the role of the Higgs boson, are often referred to as "Composite Higgs models"). Dynamical breaking of gauge symmetries is often due to creation of a fermionic condensate; for example the quark condensate, which is connected to the dynamical breaking of chiral symmetry in quantum chromodynamics. Conventional superconductivity is the paradigmatic example from the condensed matter side, where phonon-mediated attractions lead electrons to become bound in pairs and then condense, thereby breaking the electromagnetic gauge symmetry.

Generalisation and Technical Usage

For spontaneous symmetry breaking to occur, there must be a system in which there are several equally likely outcomes. The system as a whole is therefore symmetric with respect to these outcomes. (If we consider any two outcomes, the probability is the same. This contrasts sharply to explicit symmetry breaking.) However, if the system is sampled (i.e. if the system is actually used or interacted with in any way), a specific outcome must occur. Though the system as a whole is symmetric, it is never encountered with this symmetry, but only in one specific asymmetric state. Hence, the symmetry is said to be spontaneously broken in that theory. Nevertheless, the fact that each outcome is equally likely is a reflection of the underlying symmetry, which is thus often dubbed "hidden symmetry", and has crucial formal consequences.

When a theory is symmetric with respect to a symmetry group, but requires that one element of the group be distinct, then spontaneous symmetry breaking has occurred. The theory must not dictate *which* member is distinct, only that *one is*. From this point on, the theory can be treated as if this element actually is distinct, with the proviso that any results found in this way must be resymmetrized, by taking the average of each of the elements of the group being the distinct one.

The crucial concept in physics theories is the order parameter. If there is a field (often a background field) which acquires an expectation value (not necessarily a *vacuum* expectation value) which is not invariant under the symmetry in question, we say that the system is in the ordered phase, and the symmetry is spontaneously broken. This is because other subsystems interact with the order parameter, which specifies a "frame of reference" to be measured against. In that case, the vacuum state does not obey the initial symmetry (which would keep it invariant, in the linearly realized Wigner mode in which it would be a singlet), and, instead changes under the (hidden) symmetry, now implemented in the (nonlinear) Nambu–Goldstone mode. Normally, in the absence of the Higgs mechanism, massless Goldstone bosons arise.

The symmetry group can be discrete, such as the space group of a crystal, or continuous (e.g., a Lie group), such as the rotational symmetry of space. However, if the system contains only a single spatial dimension, then only discrete symmetries may be broken in a vacuum state of the full quantum theory, although a classical solution may break a continuous symmetry.

A Pedagogical Example: The Mexican Hat Potential

In the simplest idealized relativistic model, the spontaneously broken symmetry is summarized through an illustrative scalar field theory. The relevant Lagrangian, which essentially dictates how a system behaves, can be split up into kinetic and potential terms,

$$\mathcal{L} = \partial^\mu \phi \partial_\mu \phi - V(\phi). \tag{1}$$

It is in this potential term $V(\Phi)$ that the symmetry breaking is triggered. An example of a potential, due to Jeffrey Goldstone is illustrated in the graph at the right.

$$V(\phi) = -10|\phi|^2 + |\phi|^4 . \tag{2}$$

This potential has an infinite number of possible minima (vacuum states) given by

$$\phi = \sqrt{5}e^{i\theta} . \tag{3}$$

for any real θ between 0 and 2π. The system also has an unstable vacuum state corresponding to $\Phi = 0$. This state has a U(1) symmetry. However, once the system falls into a specific stable vacuum state (amounting to a choice of θ), this symmetry will appear to be lost, or "spontaneously broken".

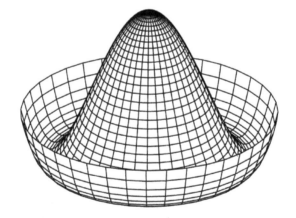

Graph of Goldstone's "Mexican hat" potential function V versus φ.

In fact, any other choice of θ would have exactly the same energy, implying the existence of a massless Nambu–Goldstone boson, the mode running around the circle at the minimum of this potential, and indicating there is some memory of the original symmetry in the Lagrangian.

Other Examples

- For ferromagnetic materials, the underlying laws are invariant under spatial rotations. Here, the order parameter is the magnetization, which measures the magnetic dipole density. Above the Curie temperature, the order parameter is zero, which is spatially invariant, and there is no symmetry breaking. Below the Curie temperature, however, the magnetization acquires a constant non-vanishing value, which points in a certain direction (in the idealized situation where we have full equilibrium; otherwise, translational symmetry gets broken as well). The residual rotational symmetries which leave the orientation of this vector invariant remain unbroken, unlike the other rotations which do not and are thus spontaneously broken.

- The laws describing a solid are invariant under the full Euclidean group, but the solid itself spontaneously breaks this group down to a space group. The displacement and the orientation are the order parameters.

- General relativity has a Lorentz symmetry, but in FRW cosmological models, the mean 4-velocity field defined by averaging over the velocities of the galaxies (the galaxies act like gas particles at cosmological scales) acts as an order parameter breaking this symmetry. Similar comments can be made about the cosmic microwave background.

- For the electroweak model, as explained earlier, a component of the Higgs field provides the order parameter breaking the electroweak gauge symmetry to the electromagnetic gauge symmetry. Like the ferromagnetic example, there is a phase transition at the electroweak temperature. The same comment about us not tending to notice broken symmetries suggests why it took so long for us to discover electroweak unification.

- In superconductors, there is a condensed-matter collective field ψ, which acts as the order parameter breaking the electromagnetic gauge symmetry.

- Take a thin cylindrical plastic rod and push both ends together. Before buckling, the system is symmetric under rotation, and so visibly cylindrically symmetric. But after buckling, it looks different, and asymmetric. Nevertheless, features of the cylindrical symmetry are still there: ignoring friction, it would take no force to freely spin the rod around, displacing the ground state in time, and amounting to an oscillation of vanishing frequency, unlike the radial oscillations in the direction of the buckle. This spinning mode is effectively the requisite Nambu–Goldstone boson.

- Consider a uniform layer of fluid over an infinite horizontal plane. This system has all the symmetries of the Euclidean plane. But now heat the bottom surface uniformly so that it becomes much hotter than the upper surface. When the temperature gradient becomes large enough, convection cells will form, breaking the Euclidean symmetry.

- Consider a bead on a circular hoop that is rotated about a vertical diameter. As the rotational velocity is increased gradually from rest, the bead will initially stay at its initial equilibrium point at the bottom of the hoop (intuitively stable, lowest gravitational potential). At a certain critical rotational velocity, this point will become unstable and the bead will jump to one of two other newly created equilibria, equidistant from the center. Initially, the system is symmetric with respect to the diameter, yet after passing the critical velocity, the bead ends up in one of the two new equilibrium points, thus breaking the symmetry.

Nobel Prize

On October 7, 2008, the Royal Swedish Academy of Sciences awarded the 2008 Nobel Prize in Physics to three scientists for their work in subatomic physics symmetry breaking. Yoichiro Nambu, of the University of Chicago, won half of the prize for the discovery of the mechanism of spontaneous broken symmetry in the context of the strong interactions, specifically chiral symmetry breaking. Physicists Makoto Kobayashi and Toshihide Maskawa shared the other half of the prize for discovering the origin of the explicit breaking of CP symmetry in the weak interactions. This origin is ultimately reliant on the Higgs mechanism, but, so far understood as a "just so" feature of Higgs couplings, not a spontaneously broken symmetry phenomenon.

References

- R. Oerter (2006). The Theory of Almost Everything: The Standard Model, the Unsung Triumph of Modern Physics (Kindle ed.). Penguin Group. p. 2. ISBN 0-13-236678-9.

- M. Gell-Mann, P. Ramond & R. Slansky (1979). F. van Nieuwenhuizen & D. Z. Freedman, eds. Supergravity. North Holland. pp. 315–321. ISBN 0-444-85438-X.

- Quantum Field Theory in a Nutshell (Second Edition), by A. Zee (Princeton University Press, 2010) ISBN 978-1-4008-3532-4.

- Mark Thomson (5 September 2013). Modern Particle Physics. Cambridge University Press. pp. 499–500. ISBN 978-1-107-29254-3.

- Close, Frank (2011). The Infinity Puzzle: Quantum Field Theory and the Hunt for an Orderly Universe. Oxford: Oxford University Press. ISBN 978-0-19-959350-7.

- I.S. Hughes (1991). "Chapter 11.1 – Cabibbo Mixing". Elementary Particles (3rd ed.). Cambridge University Press. pp. 242–243. ISBN 0-521-40402-9.

- Steven Weinberg (20 April 2011). Dreams of a Final Theory: The Scientist's Search for the Ultimate Laws of Nature. Knopf Doubleday Publishing Group. ISBN 978-0-307-78786-6.

- "Confirmed: CERN discovers new particle likely to be the Higgs boson". YouTube. Russia Today. 4 July 2012. Retrieved 2013-08-06.

- Department of Physics and Astronomy. "Rochester's Hagen Sakurai Prize Announcement". Pas.rochester.edu. Retrieved 2012-06-16.

- FermiFred (2010-02-15). "C.R. Hagen discusses naming of Higgs Boson in 2010 Sakurai Prize Talk". Youtube.com. Retrieved 2012-06-16.

5

Elementary Particles of Particle Physics

The elementary particles elucidated in the section are elementary particle, quark, photon, lepton, gluon, Higgs boson, fermion, antimatter and boson. Elementary particle is the particle whose structure is unknown and quark is the elementary particle of matter. Quark is also one of the central constitutes of matter. This text is an overview of the subject matter incorporating all the major aspects of elementary particles of particle physics.

Elementary Particle

In particle physics, an elementary particle or fundamental particle is a particle whose substructure is unknown, thus it is unknown whether it is composed of other particles. Known elementary particles include the fundamental fermions (quarks, leptons, antiquarks, and antileptons), which generally are "matter particles" and "antimatter particles", as well as the fundamental bosons (gauge bosons and the Higgs boson), which generally are "force particles" that mediate interactions among fermions. A particle containing two or more elementary particles is a *composite particle*.

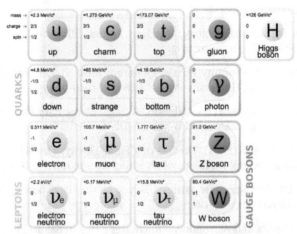

Elementary particles included in the Standard Model

Everyday matter is composed of atoms, once presumed to be matter's elementary particles—*atom* meaning "unable to cut" in Greek—although the atom's existence remained controversial until about 1910, as some leading physicists regarded molecules as mathematical illusions, and matter as ultimately composed of energy. Soon, subatomic constituents of the atom were identified. As the 1930s opened, the electron and the proton had been observed, along with the photon, the particle of electromagnetic radiation. At that time, the recent advent of quantum mechanics was radically altering the conception of particles, as a single particle could seemingly span a field as would a wave, a paradox still eluding satisfactory explanation.

Via quantum theory, protons and neutrons were found to contain quarks—up quarks and down quarks—now considered elementary particles. And within a molecule, the electron's three degrees of freedom (charge, spin, orbital) can separate via wavefunction into three quasiparticles (holon, spinon, orbiton). Yet a free electron—which, not orbiting an atomic nucleus, lacks orbital motion—appears unsplittable and remains regarded as an elementary particle.

Around 1980, an elementary particle's status as indeed elementary—an *ultimate constituent* of substance—was mostly discarded for a more practical outlook, embodied in particle physics' Standard Model, science's most experimentally successful theory. Many elaborations upon and theories beyond the Standard Model, including the extremely popular supersymmetry, double the number of elementary particles by hypothesizing that each known particle associates with a "shadow" partner far more massive, although all such superpartners remain undiscovered. Meanwhile, an elementary boson mediating gravitation—the graviton—remains hypothetical.

Overview

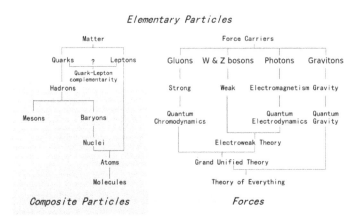

An overview of the various families of elementary and composite particles,
and the theories describing their interactions

All elementary particles are—depending on their *spin*—either bosons or fermions. These are differentiated via the spin–statistics theorem of quantum statistics. Particles of *half-integer* spin exhibit Fermi–Dirac statistics and are fermions. Particles of *integer* spin, in other words full-integer, exhibit Bose–Einstein statistics and are bosons.

Elementary Fermions:

- Matter particles

- o   Quarks:

    - • up, down

    - • charm, strange

    - • top, bottom

- o   Leptons:

    - • electron, electron neutrino (a.k.a., "neutrino")

    - • muon, muon neutrino

    - • tau, tau neutrino

- • Antimatter particles

    - o Antiquarks

    - o Antileptons

Elementary bosons:

- • Force particles (gauge bosons):

    - o photon

    - o gluon (numbering eight)

    - o W^+, W^-, and Z^0 bosons

    - o graviton (hypothetical)

- • Scalar boson

    - o Higgs boson

A particle's mass is quantified in units of energy versus the electron's (electronvolts). Through conversion of energy into mass, any particle can be produced through collision of other particles at high energy, although the output particle might not contain the input particles, for instance matter creation from colliding photons. Likewise, the composite fermions protons were collided at nearly light speed to produce the relatively more massive Higgs boson. The most massive elementary particle, the top quark, rapidly decays, but apparently does not contain, lighter particles.

When probed at energies available in experiments, particles exhibit spherical sizes. In operating particle physics' Standard Model, elementary particles are usually represented for predictive utility as point particles, which, as zero-dimensional, lack spatial extension. Though extremely successful, the Standard Model is limited to the microcosm by its omission of gravitation, and has some parameters arbitrarily added but unexplained. Seeking to resolve those shortcomings, string theory posits that elementary particles are ultimately composed of one-dimensional energy strings whose absolute minimum size is the Planck length.

Common Elementary Particles

According to the current models of big bang nucleosynthesis, the primordial composition of visible matter of the universe should be about 75% hydrogen and 25% helium-4 (in mass). Neutrons are made up of one up and two down quark, while protons are made of two up and one down quark. Since the other common elementary particles (such as electrons, neutrinos, or weak bosons) are so light or so rare when compared to atomic nuclei, we can neglect their mass contribution to the observable universe's total mass. Therefore, one can conclude that most of the visible mass of the universe consists of protons and neutrons, which, like all baryons, in turn consist of up quarks and down quarks.

Some estimates imply that there are roughly 10^{80} baryons (almost entirely protons and neutrons) in the observable universe.

The number of protons in the observable universe is called the Eddington number.

In terms of number of particles, some estimates imply that nearly all the matter, excluding dark matter, occurs in neutrinos, and that roughly 10^{86} elementary particles of matter exist in the visible universe, mostly neutrinos. Other estimates imply that roughly 10^{97} elementary particles exist in the visible universe (not including dark matter), mostly photons, gravitons, and other massless force carriers.

Standard Model

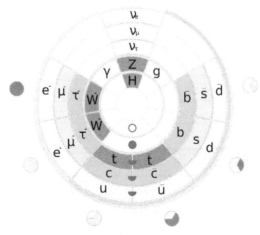

Graphic representation of the standard model. Spin, charge, mass and participation in different force interactions are shown.

The Standard Model of particle physics contains 12 flavors of elementary fermions, plus their corresponding antiparticles, as well as elementary bosons that mediate the forces and the Higgs boson, which was reported on July 4, 2012, as having been likely detected by the two main experiments at the LHC (ATLAS and CMS). However, the Standard Model is widely considered to be a provisional theory rather than a truly fundamental one, since it is not known if it is compatible with Einstein's general relativity. There may be hypothetical elementary particles not described by the Standard Model, such as the graviton, the particle that would carry the gravitational force, and sparticles, super-symmetric partners of the ordinary particles.

Fundamental Fermions

The 12 fundamental fermionic flavours are divided into three generations of four particles each. Six of the particles are quarks. The remaining six are leptons, three of which are neutrinos, and the remaining three of which have an electric charge of –1: the electron and its two cousins, the muon and the tau.

Particle Generations					
Leptons					
First generation		Second generation		Third generation	
Name	Symbol	Name	Symbol	Name	Symbol
electron	e–	muon	$\mu-$	tau	$\tau-$
electron neutrino	ν_e	muon neutrino	ν_μ	tau neutrino	ν_τ
Quarks					
First generation		Second generation		Third generation	
up quark	u	charm quark	c	top quark	t
down quark	d	strange quark	s	bottom quark	b

Antiparticles

There are also 12 fundamental fermionic antiparticles that correspond to these 12 particles. For example, the antielectron (positron) e+ is the electron's antiparticle and has an electric charge of +1.

Particle Generations					
Antileptons					
First generation		Second generation		Third generation	
Name	Symbol	Name	Symbol	Name	Symbol
positron	e+	antimuon	$\mu+$	antitau	$\tau+$
electron antineutrino	ν_e	muon antineutrino	ν_μ	tau antineutrino	ν_τ

Antiquarks					
First generation		*Second generation*		*Third generation*	
up antiquark	u	charm antiquark	c	top antiquark	t
down antiquark	d	strange antiquark	s	bottom antiquark	b

Quarks

Isolated quarks and antiquarks have never been detected, a fact explained by confinement. Every quark carries one of three color charges of the strong interaction; antiquarks similarly carry anticolor. Color-charged particles interact via gluon exchange in the same way that charged particles interact via photon exchange. However, gluons are themselves color-charged, resulting in an amplification of the strong force as color-charged particles are separated. Unlike the electromagnetic force, which diminishes as charged particles separate, color-charged particles feel increasing force.

However, color-charged particles may combine to form color neutral composite particles called hadrons. A quark may pair up with an antiquark: the quark has a color and the antiquark has the corresponding anticolor. The color and anticolor cancel out, forming a color neutral meson. Alternatively, three quarks can exist together, one quark being "red", another "blue", another "green". These three colored quarks together form a color-neutral baryon. Symmetrically, three antiquarks with the colors "antired", "antiblue" and "antigreen" can form a color-neutral antibaryon.

Quarks also carry fractional electric charges, but, since they are confined within hadrons whose charges are all integral, fractional charges have never been isolated. Note that quarks have electric charges of either +2/3 or −1/3, whereas antiquarks have corresponding electric charges of either −2/3 or +1/3.

Evidence for the existence of quarks comes from deep inelastic scattering: firing electrons at nuclei to determine the distribution of charge within nucleons (which are baryons). If the charge is uniform, the electric field around the proton should be uniform and the electron should scatter elastically. Low-energy electrons do scatter in this way, but, above a particular energy, the protons deflect some electrons through large angles. The recoiling electron has much less energy and a jet of particles is emitted. This inelastic scattering suggests that the charge in the proton is not uniform but split among smaller charged particles: quarks.

Fundamental Bosons

In the Standard Model, vector (spin-1) bosons (gluons, photons, and the W and Z bosons) mediate forces, whereas the Higgs boson (spin-0) is responsible for the intrinsic mass of particles. Bosons differ from fermions in the fact that multiple bosons can occupy the same quantum state (Pauli exclusion principle). Also, bosons can be either elementary, like photons, or a combination, like mesons. The spin of bosons are integers instead of half integers.

Gluons

Gluons mediate the strong interaction, which join quarks and thereby form hadrons, which are either baryons (three quarks) or mesons (one quark and one antiquark). Protons and neutrons are baryons, joined by gluons to form the atomic nucleus. Like quarks, gluons exhibit colour and anticolour—unrelated to the concept of visual color— sometimes in combinations, altogether eight variations of gluons.

Electroweak Bosons

There are three weak gauge bosons: W^+, W^-, and Z^o; these mediate the weak interaction. The W bosons are known for their mediation in nuclear decay. The W^- converts a neutron into a proton then decay into an electron and electron antineutrino pair. The Z^o does not convert charge but rather changes momentum and is the only mechanism for elastically scattering neutrinos. The weak gauge bosons were discovered due to momentum change in electrons from neutrino-Z exchange. The massless photon mediates the electromagnetic interaction. These four gauge bosons form the electroweak interaction among elementary particles.

Higgs Boson

Although the weak and electromagnetic forces appear quite different to us at everyday energies, the two forces are theorized to unify as a single electroweak force at high energies. This prediction was clearly confirmed by measurements of cross-sections for high-energy electron-proton scattering at the HERA collider at DESY. The differences at low energies is a consequence of the high masses of the W and Z bosons, which in turn are a consequence of the Higgs mechanism. Through the process of spontaneous symmetry breaking, the Higgs selects a special direction in electroweak space that causes three electroweak particles to become very heavy (the weak bosons) and one to remain massless (the photon). On 4 July 2012, after many years of experimentally searching for evidence of its existence, the Higgs boson was announced to have been observed at CERN's Large Hadron Collider. Peter Higgs who first posited the existence of the Higgs boson was present at the announcement. The Higgs boson is believed to have a mass of approximately 125 GeV. The statistical significance of this discovery was reported as 5-sigma, which implies a certainty of roughly 99.99994%. In particle physics, this is the level of significance required to officially label experimental observations as a discovery. Research into the properties of the newly discovered particle continues.

Graviton

The graviton is hypothesized to mediate gravitation, but remains undiscovered and yet is sometimes included in tables of elementary particles. Its spin would be two—thus a boson—and it would lack charge or mass. Besides mediating an extremely feeble force, the graviton would have its own antiparticle and rapidly annihilate, rendering its detection extremely difficult even if it exists.

Beyond the Standard Model

Although experimental evidence overwhelmingly confirms the predictions derived from the Standard Model, some of its parameters were added arbitrarily, not determined by a particular explanation, which remain mysteries, for instance the hierarchy problem. Theories beyond the Standard Model attempt to resolve these shortcomings.

Grand Unification

One extension of the Standard Model attempts to combine the electroweak interaction with the strong interaction into a single 'grand unified theory' (GUT). Such a force would be spontaneously broken into the three forces by a Higgs-like mechanism. The most dramatic prediction of grand unification is the existence of X and Y bosons, which cause proton decay. However, the non-observation of proton decay at the Super-Kamiokande neutrino observatory rules out the simplest GUTs, including SU(5) and SO(10).

Supersymmetry

Supersymmetry extends the Standard Model by adding another class of symmetries to the Lagrangian. These symmetries exchange fermionic particles with bosonic ones. Such a symmetry predicts the existence of supersymmetric particles, abbreviated as *sparticles*, which include the sleptons, squarks, neutralinos, and charginos. Each particle in the Standard Model would have a superpartner whose spin differs by 1/2 from the ordinary particle. Due to the breaking of supersymmetry, the sparticles are much heavier than their ordinary counterparts; they are so heavy that existing particle colliders would not be powerful enough to produce them. However, some physicists believe that sparticles will be detected by the Large Hadron Collider at CERN.

String Theory

String theory is a model of physics where all "particles" that make up matter are composed of strings (measuring at the Planck length) that exist in an 11-dimensional (according to M-theory, the leading version) universe. These strings vibrate at different frequencies that determine mass, electric charge, color charge, and spin. A string can be open (a line) or closed in a loop (a one-dimensional sphere, like a circle). As a string moves through space it sweeps out something called a *world sheet*. String theory predicts 1- to 10-branes (a 1-brane being a string and a 10-brane being a 10-dimensional object) that prevent tears in the "fabric" of space using the uncertainty principle (E.g., the electron orbiting a hydrogen atom has the probability, albeit small, that it could be anywhere else in the universe at any given moment).

String theory proposes that our universe is merely a 4-brane, inside which exist the 3 space dimensions and the 1 time dimension that we observe. The remaining 6 theoretical dimensions either are very tiny and curled up (and too small to be macroscopically accessible) or simply do not/cannot exist in our universe (because they exist in a grander scheme called the "multiverse" outside our known universe).

Some predictions of the string theory include existence of extremely massive counterparts of ordinary particles due to vibrational excitations of the fundamental string and existence of a massless spin-2 particle behaving like the graviton.

Technicolor

Technicolor theories try to modify the Standard Model in a minimal way by introducing a new QCD-like interaction. This means one adds a new theory of so-called Techniquarks, interacting via so called Technigluons. The main idea is that the Higgs-Boson is not an elementary particle but a bound state of these objects.

Preon Theory

According to preon theory there are one or more orders of particles more fundamental than those (or most of those) found in the Standard Model. The most fundamental of these are normally called preons, which is derived from "pre-quarks". In essence, preon theory tries to do for the Standard Model what the Standard Model did for the particle zoo that came before it. Most models assume that almost everything in the Standard Model can be explained in terms of three to half a dozen more fundamental particles and the rules that govern their interactions. Interest in preons has waned since the simplest models were experimentally ruled out in the 1980s.

Acceleron Theory

Accelerons are the hypothetical subatomic particles that integrally link the newfound mass of the neutrino to the dark energy conjectured to be accelerating the expansion of the universe.

In theory, neutrinos are influenced by a new force resulting from their interactions with accelerons. Dark energy results as the universe tries to pull neutrinos apart.

Quark

A quark (/ˈkwɔːrk/ or /ˈkwɑːrk/) is an elementary particle and a fundamental constituent of matter. Quarks combine to form composite particles called hadrons, the most stable of which are protons and neutrons, the components of atomic nuclei. Due to a phenomenon known as *color confinement*, quarks are never directly observed or found in isolation; they can be found only within hadrons, such as baryons (of which protons and neutrons are examples), and mesons. For this reason, much of what is known about quarks has been drawn from observations of the hadrons themselves.

Quarks have various intrinsic properties, including electric charge, mass, color charge and spin. Quarks are the only elementary particles in the Standard Model of particle physics to experience all four fundamental interactions, also known as *fundamental forces* (electromagnetism, gravitation, strong interaction, and weak interaction), as well as the only known particles whose electric charges are not integer multiples of the elementary charge.

There are six types of quarks, known as *flavors*: up, down, strange, charm, top, and bottom. Up and down quarks have the lowest masses of all quarks. The heavier quarks rapidly change into up and down quarks through a process of particle decay: the transformation from a higher mass state to a lower mass state. Because of this, up and down quarks are generally stable and the most common in the universe, whereas strange, charm, bottom, and top quarks can only be produced in high energy collisions (such as those involving cosmic rays and in particle accelerators). For every quark flavor there is a corresponding type of antiparticle, known as an *antiquark*, that differs from the quark only in that some of its properties have equal magnitude but opposite sign.

The quark model was independently proposed by physicists Murray Gell-Mann and George Zweig in 1964. Quarks were introduced as parts of an ordering scheme for hadrons, and there was little evidence for their physical existence until deep inelastic scattering experiments at the Stanford Linear Accelerator Center in 1968. Accelerator experiments have provided evidence for all six flavors. The top quark was the last to be discovered at Fermilab in 1995.

Classification

The Standard Model is the theoretical framework describing all the currently known elementary particles. This model contains six flavors of quarks (q), named up (u), down (d), strange (s), charm (c), bottom (b), and top (t). Antiparticles of quarks are called *antiquarks*, and are denoted by a bar over the symbol for the corresponding quark, such as u for an up antiquark. As with antimatter in general, antiquarks have the same mass, mean lifetime, and spin as their respective quarks, but the electric charge and other charges have the opposite sign.

Quarks are spin-$\frac{1}{2}$ particles, implying that they are fermions according to the spin-statistics theorem. They are subject to the Pauli exclusion principle, which states that no two identical fermions can simultaneously occupy the same quantum state. This is in contrast to bosons (particles with integer spin), any number of which can be in the same state. Unlike leptons, quarks possess color charge, which causes them to engage in the strong interaction. The resulting attraction between different quarks causes the formation of composite particles known as *hadrons*.

The quarks which determine the quantum numbers of hadrons are called *valence quarks*; apart from these, any hadron may contain an indefinite number of virtual (or *sea*) quarks, antiquarks, and gluons which do not influence its quantum numbers. There are two families of hadrons: baryons, with three valence quarks, and mesons, with a valence quark and an antiquark. The most common baryons are the proton and the neutron, the building blocks of the atomic nucleus. A great number of hadrons are known, most of them differentiated by their quark content and the properties these constituent quarks confer. The existence of "exotic" hadrons with more valence quarks, such as tetraquarks (qqqq) and pentaquarks (qqqqq), has been conjectured but not proven. However, on 13 July 2015, the LHCb collaboration at CERN reported results consistent with pentaquark states.

Elementary fermions are grouped into three generations, each comprising two leptons and two quarks. The first generation includes up and down quarks, the second strange and charm quarks, and the third bottom and top quarks. All searches for a fourth generation of quarks and other elementary fermions have failed, and there is strong indirect evidence that no more than three generations exist.[nb 2] Particles in higher generations generally have greater mass and less stability, causing them to decay into lower-generation particles by means of weak interactions. Only first-generation (up and down) quarks occur commonly in nature. Heavier quarks can only be created in high-energy collisions (such as in those involving cosmic rays), and decay quickly; however, they are thought to have been present during the first fractions of a second after the Big Bang, when the universe was in an extremely hot and dense phase (the quark epoch). Studies of heavier quarks are conducted in artificially created conditions, such as in particle accelerators.

Having electric charge, mass, color charge, and flavor, quarks are the only known elementary particles that engage in all four fundamental interactions of contemporary physics: electromagnetism, gravitation, strong interaction, and weak interaction. Gravitation is too weak to be relevant to individual particle interactions except at extremes of energy (Planck energy) and distance scales (Planck distance). However, since no successful quantum theory of gravity exists, gravitation is not described by the Standard Model.

History

The quark model was independently proposed by physicists Murray Gell-Mann (pictured) and George Zweig in 1964. The proposal came shortly after Gell-Mann's 1961 formulation of a particle classification system known as the *Eightfold Way*—or, in more technical terms, SU(3) flavor symmetry. Physicist Yuval Ne'eman had independently developed a scheme similar to the Eightfold Way in the same year.

Murray Gell-Mann at TED in 2007. Gell-Mann and George Zweig proposed the quark model in 1964.

At the time of the quark theory's inception, the "particle zoo" included, amongst other particles, a multitude of hadrons. Gell-Mann and Zweig posited that they were not elementary particles, but were instead composed of combinations of quarks and antiquarks. Their model involved three flavors of quarks, up, down, and strange, to which they ascribed properties such as spin and electric charge. The initial reaction of the physics community to the proposal was mixed. There was particular contention about whether the quark was a physical entity or a mere abstraction used to explain concepts that were not fully understood at the time.

In less than a year, extensions to the Gell-Mann–Zweig model were proposed. Sheldon Lee Glashow and James Bjorken predicted the existence of a fourth flavor of quark, which they called *charm*. The addition was proposed because it allowed for a better description of the weak interaction (the mechanism that allows quarks to decay), equalized the number of known quarks with the number of known leptons, and implied a mass formula that correctly reproduced the masses of the known mesons.

In 1968, deep inelastic scattering experiments at the Stanford Linear Accelerator Center (SLAC) showed that the proton contained much smaller, point-like objects and was therefore not an elementary particle. Physicists were reluctant to firmly identify these objects with quarks at the time, instead calling them "partons"—a term coined by Richard Feynman. The objects that were observed at SLAC would later be identified as up and down quarks as the other flavors were discovered. Nevertheless, "parton" remains in use as a collective term for the constituents of hadrons (quarks, antiquarks, and gluons).

The strange quark's existence was indirectly validated by SLAC's scattering experiments: not only was it a necessary component of Gell-Mann and Zweig's three-quark model, but it provided an explanation for the kaon (K) and pion (π) hadrons discovered in cosmic rays in 1947.

In a 1970 paper, Glashow, John Iliopoulos and Luciano Maiani presented the so-called GIM mechanism to explain the experimental non-observation of flavor-changing neutral currents. This theoretical model required the existence of the as-yet undiscovered charm quark. The number of supposed quark flavors grew to the current six in 1973, when Makoto Kobayashi and Toshihide Maskawa noted that the experimental observation of CP violation[nb 3] could be explained if there were another pair of quarks.

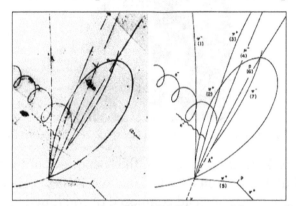

Photograph of the event that led to the discovery of the Σ++c baryon,
at the Brookhaven National Laboratory in 1974

Charm quarks were produced almost simultaneously by two teams in November 1974 (November Revolution)—one at SLAC under Burton Richter, and one at Brookha-ven National Laboratory under Samuel Ting. The charm quarks were observed bound with charm antiquarks in mesons. The two parties had assigned the discovered meson two different symbols, J and ψ; thus, it became formally known as the J/ψ meson. The discovery finally convinced the physics community of the quark model's validity.

In the following years a number of suggestions appeared for extending the quark model to six quarks. Of these, the 1975 paper by Haim Harari was the first to coin the terms *top* and *bottom* for the additional quarks.

In 1977, the bottom quark was observed by a team at Fermilab led by Leon Lederman. This was a strong indicator of the top quark's existence: without the top quark, the bottom quark would have been without a partner. However, it was not until 1995 that the top quark was finally observed, also by the CDF and DØ teams at Fermilab. It had a mass much larger than had been previously expected, almost as large as that of a gold atom.

Etymology

For some time, Gell-Mann was undecided on an actual spelling for the term he intended to coin, until he found the word *quark* in James Joyce's book *Finnegans Wake*:

Three quarks for Muster Mark! Sure he has not got much of a bark And sure any he has it's all beside the mark.

—James Joyce, Finnegans Wake

Gell-Mann went into further detail regarding the name of the quark in his book *The Quark and the Jaguar*:

In 1963, when I assigned the name "quark" to the fundamental constituents of the nucleon, I had the sound first, without the spelling, which could have been "kwork". Then, in one of my occasional perusals of *Finnegans Wake*, by James Joyce, I came across the word "quark" in the phrase "Three quarks for Muster Mark". Since "quark" (meaning, for one thing, the cry of the gull) was clearly intended to rhyme with "Mark", as well as "bark" and other such words, I had to find an excuse to pronounce it as "kwork". But the book represents the dream of a publican named Humphrey Chimpden Earwicker. Words in the text are typically drawn from several sources at once, like the "portmanteau" words in *Through the Looking-Glass*. From time to time, phrases occur in the book that are partially determined by calls for drinks at the bar. I argued, therefore, that perhaps one of the multiple sources of the cry "Three quarks for Muster Mark" might be "Three quarts for Mister Mark", in which case the pronunciation "kwork" would not be totally unjustified. In any case, the number three fitted perfectly the way quarks occur in nature.

Zweig preferred the name *ace* for the particle he had theorized, but Gell-Mann's terminology came to prominence once the quark model had been commonly accepted.

The quark flavors were given their names for several reasons. The up and down quarks are named after the up and down components of isospin, which they carry. Strange quarks were given their name because they were discovered to be components of the strange particles discovered in cosmic rays years before the quark model was proposed; these particles were deemed "strange" because they had unusually long lifetimes. Glashow, who coproposed charm quark with Bjorken, is quoted as saying, "We called our construct the 'charmed quark', for we were fascinated and pleased by the symmetry it brought to the subnuclear world." The names "bottom" and "top", coined by Harari, were chosen because they are "logical partners for up and down quarks". In the past, bottom and top quarks were sometimes referred to as "beauty" and "truth" respectively, but these names have somewhat fallen out of use. While "truth" never did catch on, accelerator complexes devoted to massive production of bottom quarks are sometimes called "beauty factories".

Properties

Electric Charge

Quarks have fractional electric charge values – either $\frac{1}{3}$ or $\frac{2}{3}$ times the elementary charge (e), depending on flavor. Up, charm, and top quarks (collectively referred to as *up-type quarks*) have a charge of $+\frac{2}{3}$ e, while down, strange, and bottom quarks (*down-type quarks*) have $-\frac{1}{3}$ e. Antiquarks have the opposite charge to their corresponding quarks; up-type antiquarks have charges of $-\frac{2}{3}$ e and down-type antiquarks have charges of $+\frac{1}{3}$ e. Since the electric charge of a hadron is the sum of the charges of the constituent quarks, all hadrons have integer charges: the combination of three quarks (baryons), three antiquarks (antibaryons), or a quark and an antiquark (mesons) always results in integer charges. For example, the hadron constituents of atomic nuclei, neutrons and protons, have charges of 0 e and +1 e respectively; the neutron is composed of two down quarks and one up quark, and the proton of two up quarks and one down quark.

Spin

Spin is an intrinsic property of elementary particles, and its direction is an important degree of freedom. It is sometimes visualized as the rotation of an object around its own axis (hence the name "spin"), though this notion is somewhat misguided at subatomic scales because elementary particles are believed to be point-like.

Spin can be represented by a vector whose length is measured in units of the reduced Planck constant $\hbar$ (pronounced "h bar"). For quarks, a measurement of the spin vector component along any axis can only yield the values $+\hbar/2$ or $-\hbar/2$; for this reason quarks are classified as spin-$\frac{1}{2}$ particles. The component of spin along a given axis – by convention the z axis – is often denoted by an up arrow ↑ for the value $+\frac{1}{2}$ and down arrow ↓ for the value $-\frac{1}{2}$, placed after the symbol for flavor. For example, an up quark with a spin of $+\frac{1}{2}$ along the z axis is denoted by u↑.

Weak Interaction

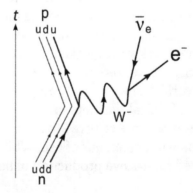

Feynman diagram of beta decay with time flowing upwards. The CKM matrix (discussed below) encodes the probability of this and other quark decays.

A quark of one flavor can transform into a quark of another flavor only through the weak interaction, one of the four fundamental interactions in particle physics. By absorbing or emitting a W boson, any up-type quark (up, charm, and top quarks) can change into any down-type quark (down, strange, and bottom quarks) and vice versa. This flavor transformation mechanism causes the radioactive process of beta decay, in which a neutron (n) "splits" into a proton (p), an electron (e−) and an electron antineutrino (v e) (see picture). This occurs when one of the down quarks in the neutron (udd) decays into an up quark by emitting a virtual W− boson, transforming the neutron into a proton (uud). The W− boson then decays into an electron and an electron antineutrino.

| n | → | p | + | e− | + | v_e | (Beta decay, hadron notation) |
| udd | → | uud | + | e− | + | v_e | (Beta decay, quark notation) |

Both beta decay and the inverse process of *inverse beta decay* are routinely used in medical applications such as positron emission tomography (PET) and in experiments involving neutrino detection.

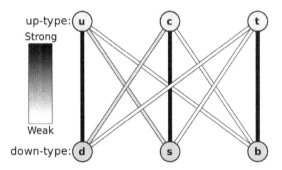

The strengths of the weak interactions between the six quarks. The "intensities" of the lines are determined by the elements of the CKM matrix.

While the process of flavor transformation is the same for all quarks, each quark has a preference to transform into the quark of its own generation. The relative tendencies of all flavor transformations are described by a mathematical table, called the Cabibbo–Kobayashi–Maskawa matrix (CKM matrix). Enforcing unitarity, the approximate magnitudes of the entries of the CKM matrix are:

$$
\begin{bmatrix} |V_{ud}| & |V_{us}| & |V_{ub}| \\ |V_{cd}| & |V_{cs}| & |V_{cb}| \\ |V_{td}| & |V_{ts}| & |V_{tb}| \end{bmatrix} \approx \begin{bmatrix} 0.974 & 0.225 & 0.003 \\ 0.225 & 0.973 & 0.041 \\ 0.009 & 0.040 & 0.999 \end{bmatrix},
$$

where V_{ij} represents the tendency of a quark of flavor i to change into a quark of flavor j (or vice versa).

There exists an equivalent weak interaction matrix for leptons (right side of the W boson on the above beta decay diagram), called the Pontecorvo–Maki–Nakagawa–Sakata matrix (PMNS matrix). Together, the CKM and PMNS matrices describe all flavor transformations, but the links between the two are not yet clear.

Strong Interaction and Color Charge

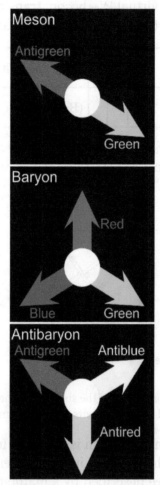

All types of hadrons have zero total color charge.

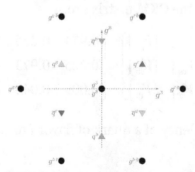

The pattern of strong charges for the three colors of quark, three antiquarks,
and eight gluons (with two of zero charge overlapping).

According to quantum chromodynamics (QCD), quarks possess a property called *color charge*. There are three types of color charge, arbitrarily labeled *blue, green,* and *red*. Each of them is complemented by an anticolor – *antiblue, antigreen,* and *antired*. Every quark carries a color, while every antiquark carries an anticolor.

The system of attraction and repulsion between quarks charged with different combinations of the three colors is called strong interaction, which is mediated by force carrying particles known as *gluons*; this is discussed at length below. The theory that describes strong interactions is called quantum chromodynamics (QCD). A quark, which will have a single color value, can form a bound system with an antiquark carrying the corresponding anticolor. The result of two attracting quarks will be color neutrality: a quark with color charge ξ plus an antiquark with color charge $-\xi$ will result in a color charge of 0 (or "white" color) and the formation of a meson. This is analogous to the additive color model in basic optics. Similarly, the combination of three quarks, each with different color charges, or three antiquarks, each with anticolor charges, will result in the same "white" color charge and the formation of a baryon or antibaryon.

In modern particle physics, gauge symmetries – a kind of symmetry group – relate interactions between particles. Color SU(3) (commonly abbrevi-ated to SU(3)$_c$) is the gauge symmetry that relates the color charge in quarks and is the defining symmetry for quantum chromodynamics. Just as the laws of physics are independent of which directions in space are designated x, y, and z, and remain un-changed if the coordinate axes are rotated to a new orientation, the physics of quantum chromodynamics is independent of which directions in three-dimensional color space are identified as blue, red, and green. SU(3)$_c$ color transformations correspond to "rotations" in color space (which, mathematically speaking, is a complex space). Every quark flavor f, each with subtypes f_B, f_G, f_R corresponding to the quark colors, forms a triplet: a three-component quantum field which transforms under the fundamental representation of SU(3)$_c$. The requirement that SU(3)$_c$ should be local – that is, that its transformations be allowed to vary with space and time – determines the properties of the strong interaction. In particular, it implies the existence of eight gluon types to act as its force carriers.

Mass

Two terms are used in referring to a quark's mass: *current quark mass* refers to the mass of a quark by itself, while *constituent quark mass* refers to the current quark mass plus the mass of the gluon particle field surrounding the quark. These masses typically have very different values. Most of a hadron's mass comes from the gluons that bind the constituent quarks together, rather than from the quarks themselves. While gluons are inherently massless, they possess energy – more specifically, quantum chromodynamics binding energy (QCBE) – and it is this that contributes so greatly to the overall mass of the hadron (see mass in special relativity). For example, a proton has a mass of approximately 938 MeV/c², of which the rest mass of its three valence quarks only contributes about 9 MeV/c²; much of the remainder can be attributed to the field energy of the gluons.

Current quark masses for all six flavors in comparison, as balls of proportional volumes. Proton and electron (red) are shown in bottom left corner for scale

The Standard Model posits that elementary particles derive their masses from the Higgs mechanism, which is associated to the Higgs boson. It is hoped that further research into the reasons for the top quark's large mass of ~173 GeV/c², almost the mass of a gold atom, might reveal more about the origin of the mass of quarks and other elementary particles.

Table of Properties

The following table summarizes the key properties of the six quarks. Flavor quantum numbers (isospin (I_3), charm (C), strangeness (S), topness (T), and bottomness (B')) are assigned to certain quark flavors, and denote qualities of quark-based systems and hadrons. The baryon number (B) is $+\frac{1}{3}$ for all quarks, as baryons are made of three quarks. For antiquarks, the electric charge (Q) and all flavor quantum numbers (B, I_3, C, S, T, and B') are of opposite sign. Mass and total angular momentum (J; equal to spin for point particles) do not change sign for the antiquarks.

Quark flavor properties												
Name	Sym-bol	Mass (MeV/c²)*	J	B	Q (e)	I_3	C	S	T	B'	Antipar-ticle	Antiparti-cle symbol
First generation												
Up	u	2.3±0.7 ± 0.5	$\frac{1}{2}$	$+\frac{1}{3}$	$+\frac{2}{3}$	$+\frac{1}{2}$	0	0	0	0	Antiup	u
Down	d	4.8±0.5 ± 0.3	$\frac{1}{2}$	$+\frac{1}{3}$	$-\frac{1}{3}$	$-\frac{1}{2}$	0	0	0	0	Antidown	d
Second generation												
Charm	c	1275±25	$\frac{1}{2}$	$+\frac{1}{3}$	$+\frac{2}{3}$	0	+1	0	0	0	Antich-arm	c
Strange	s	95±5	$\frac{1}{2}$	$+\frac{1}{3}$	$-\frac{1}{3}$	0	0	-1	0	0	An-tistrange	s

Third generation												
Top	t	173210±510 ± 710	$\frac{1}{2}$	$+\frac{1}{3}$	$+\frac{2}{3}$	0	0	0	+1	0	Antitop	t
Bottom	b	4180±30	$\frac{1}{2}$	$+\frac{1}{3}$	$-\frac{1}{3}$	0	0	0	0	−1	Antibot-tom	b

J = total angular momentum, B = baryon number, Q = electric charge, I_3 = isospin, C = charm, S = strangeness, T = topness, B' = bottomness. * Notation such as 173210±510 ± 710 denotes two types of measurement uncertainty. In the case of the top quark, the first uncertainty is statistical in nature, and the second is systematic.

Interacting Quarks

As described by quantum chromodynamics, the strong interaction between quarks is mediated by gluons, massless vector gauge bosons. Each gluon carries one color charge and one anticolor charge. In the standard framework of particle interactions (part of a more general formulation known as perturbation theory), gluons are constantly exchanged between quarks through a virtual emission and absorption process. When a gluon is transferred between quarks, a color change occurs in both; for example, if a red quark emits a red–antigreen gluon, it becomes green, and if a green quark absorbs a red–antigreen gluon, it becomes red. Therefore, while each quark's color constantly changes, their strong interaction is preserved.

Since gluons carry color charge, they themselves are able to emit and absorb other gluons. This causes *asymptotic freedom*: as quarks come closer to each other, the chromodynamic binding force between them weakens. Conversely, as the distance between quarks increases, the binding force strengthens. The color field becomes stressed, much as an elastic band is stressed when stretched, and more gluons of appropriate color are spontaneously created to strengthen the field. Above a certain energy threshold, pairs of quarks and antiquarks are created. These pairs bind with the quarks being separated, causing new hadrons to form. This phenomenon is known as *color confinement*: quarks never appear in isolation. This process of hadronization occurs before quarks, formed in a high energy collision, are able to interact in any other way. The only exception is the top quark, which may decay before it hadronizes.

Sea Quarks

Hadrons contain, along with the *valence quarks* (qv) that contribute to their quantum numbers, virtual quark–antiquark (qq) pairs known as *sea quarks* (qs). Sea quarks form when a gluon of the hadron's color field splits; this process also works in reverse in that the annihilation of two sea quarks produces a gluon. The result is a constant flux of gluon splits and creations colloquially known as "the sea". Sea quarks are much less stable than their valence counterparts, and they typically annihilate each other within the interior of the hadron. Despite this, sea quarks can hadronize into baryonic or mesonic particles under certain circumstances.

Other Phases of Quark Matter

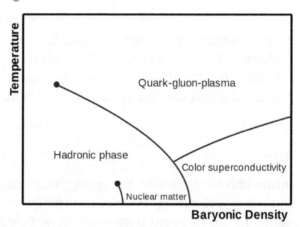

A qualitative rendering of the phase diagram of quark matter. The precise details of the diagram are the subject of ongoing research.

Under sufficiently extreme conditions, quarks may become deconfined and exist as free particles. In the course of asymptotic freedom, the strong interaction becomes weaker at higher temperatures. Eventually, color confinement would be lost and an extremely hot plasma of freely moving quarks and gluons would be formed. This theoretical phase of matter is called quark–gluon plasma. The exact conditions needed to give rise to this state are unknown and have been the subject of a great deal of speculation and experimentation. A recent estimate puts the needed temperature at $(1.90\pm0.02)\times10^{12}$ kelvin. While a state of entirely free quarks and gluons has never been achieved (despite numerous attempts by CERN in the 1980s and 1990s), recent experiments at the Relativistic Heavy Ion Collider have yielded evidence for liquid-like quark matter exhibiting "nearly perfect" fluid motion.

The quark–gluon plasma would be characterized by a great increase in the number of heavier quark pairs in relation to the number of up and down quark pairs. It is believed that in the period prior to 10^{-6} seconds after the Big Bang (the quark epoch), the universe was filled with quark–gluon plasma, as the temperature was too high for hadrons to be stable.

Given sufficiently high baryon densities and relatively low temperatures – possibly comparable to those found in neutron stars – quark matter is expected to degenerate into a Fermi liquid of weakly interacting quarks. This liquid would be characterized by a condensation of colored quark Cooper pairs, thereby breaking the local $SU(3)_c$ symmetry. Because quark Cooper pairs harbor color charge, such a phase of quark matter would be color superconductive; that is, color charge would be able to pass through it with no resistance.

Photon

A photon is an elementary particle, the quantum of all forms of electromagnetic radiation including light. It is the force carrier for electromagnetic force, even when static via virtual photons. The photon has zero rest mass and as a result, the interactions of this force with matter at long distance are observable at the microscopic and macroscopic levels. Like all elementary particles, photons are currently best explained by quantum mechanics but exhibit wave–particle duality, exhibiting properties of both waves and particles. For example, a single photon may be refracted by a lens and exhibit wave interference with itself, and it can behave as a particle with definite and finite measurable position and momentum. The photon's wave and quanta qualities are two observable aspects of a single phenomenon, and cannot be described by any mechanical model; a representation of this dual property of light, which assumes certain points on the wavefront to be the seat of the energy, is not possible. The quanta in a light wave cannot be spatially localized. Some defined physical parameters of a photon are listed.

The modern concept of the photon was developed gradually by Albert Einstein in the early 20th century to explain experimental observations that did not fit the classical wave model of light. The benefit of the photon model was that it accounted for the frequency dependence of light's energy, and explained the ability of matter and electromagnetic radiation to be in thermal equilibrium. The photon model accounted for anomalous observations, including the properties of black-body radiation, that others (notably Max Planck) had tried to explain using *semiclassical models*. In that model, light was described by Maxwell's equations, but material objects emitted and absorbed light in *quantized* amounts (i.e., they change energy only by certain particular discrete amounts). Although these semiclassical models contributed to the development of quantum mechanics, many further experiments beginning with the phenomenon of Compton scattering of single photons by electrons, validated Einstein's hypothesis that *light itself* is quantized. In 1926 the optical physicist Frithiof Wolfers and the chemist Gilbert N. Lewis coined the name photon for these particles. After Arthur H. Compton won the Nobel Prize in 1927 for his scattering studies, most scientists accepted that light quanta have an independent existence, and the term *photon* was accepted.

In the Standard Model of particle physics, photons and other elementary particles are described as a necessary consequence of physical laws having a certain symmetry at every point in spacetime. The intrinsic properties of particles, such as charge, mass and spin, are determined by this gauge symmetry. The photon concept has led to momentous advances in experimental and theoretical physics, including lasers, Bose–Einstein condensation, quantum field theory, and the probabilistic interpretation of quantum mechanics. It has been applied to photochemistry, high-resolution microscopy, and measurements of molecular distances. Recently, photons have been studied as elements of quantum computers, and for applications in optical imaging and optical communication such as quantum cryptography.

Nomenclature

In 1900, the German physicist Max Planck was studying black-body radiation and suggested that the energy carried by electromagnetic waves could only be released in "packets" of energy. In his 1901 article in *Annalen der Physik* he called these packets "energy elements". The word *quanta* (singular *quantum,* Latin for *how much*) was used before 1900 to mean particles or amounts of different quantities, including electricity. In 1905, Albert Einstein suggested that electromagnetic waves could only exist as discrete wave-packets. He called such a wave-packet *the light quantum* (German: *das Lichtquant*).[Note 1] The name *photon* derives from the Greek word for light, φ□ς (transliterated *phôs*). Arthur Compton used *photon* in 1928, referring to Gilbert N. Lewis. The same name was used earlier, by the American physicist and psychologist Leonard T. Troland, who coined the word in 1916, in 1921 by the Irish physicist John Joly, in 1924 by the French physiologist René Wurmser (1890-1993) and in 1926 by the French physicist Frithiof Wolfers (1891-1971). The name was suggested initially as a unit related to the illumination of the eye and the resulting sensation of light and was used later in a physiological context. Although Wolfers's and Lewis's theories were contradicted by many experiments and never accepted, the new name was adopted very soon by most physicists after Compton used it.

In physics, a photon is usually denoted by the symbol γ (the Greek letter gamma). This symbol for the photon probably derives from gamma rays, which were discovered in 1900 by Paul Villard, named by Ernest Rutherford in 1903, and shown to be a form of electromagnetic radiation in 1914 by Rutherford and Edward Andrade. In chemistry and optical engineering, photons are usually symbolized by $h\nu$, the energy of a photon, where h is Planck's constant and the Greek letter ν (nu) is the photon's frequency. Much less commonly, the photon can be symbolized by hf, where its frequency is denoted by f.

Physical Properties

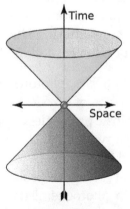

The cone shows possible values of wave 4-vector of a photon. The "time" axis gives the angular frequency (rad·s⁻¹) and the "space" axes represent the angular wavenumber (rad·m⁻¹). Green and indigo represent left and right polarization

A photon is massless, has no electric charge, and is a stable particle. A photon has two possible polarization states. In the momentum representation of the photon, which is preferred in quantum field theory, a photon is described by its wave vector, which determines its wavelength λ and its direction of propagation. A photon's wave vector may not be zero and can be represented either as a spatial 3-vector or as a (relativistic) four-vector; in the latter case it belongs to the light cone (pictured). Different signs of the four-vector denote different circular polarizations, but in the 3-vector representation one should account for the polarization state separately; it actually is a spin quantum number. In both cases the space of possible wave vectors is three-dimensional.

The photon is the gauge boson for electromagnetism, and therefore all other quantum numbers of the photon (such as lepton number, baryon number, and flavour quantum numbers) are zero. Also, the photon does not obey the Pauli exclusion principle.

Photons are emitted in many natural processes. For example, when a charge is accelerated it emits synchrotron radiation. During a molecular, atomic or nuclear transition to a lower energy level, photons of various energy will be emitted, ranging from radio waves to gamma rays. A photon can also be emitted when a particle and its corresponding antiparticle are annihilated (for example, electron–positron annihilation).

In empty space, the photon moves at c (the speed of light) and its energy and momentum are related by $E = pc$, where p is the magnitude of the momentum vector p. This derives from the following relativistic relation, with $m = 0$:

$$E^2 = p^2 c^2 + m^2 c^4.$$

The energy and momentum of a photon depend only on its frequency (ν) or inversely, its wavelength (λ):

$$E = \hbar\omega = h\nu = \frac{hc}{\lambda}$$

$$\mathbf{p} = \hbar\mathbf{k},$$

where k is the wave vector (where the wave number $k = |k| = 2\pi/\lambda$), $\omega = 2\pi\nu$ is the angular frequency, and $\hbar = h/2\pi$ is the reduced Planck constant.

Since p points in the direction of the photon's propagation, the magnitude of the momentum is

$$p = \hbar k = \frac{h\nu}{c} = \frac{h}{\lambda}.$$

The photon also carries a quantity called spin angular momentum that does not depend on its frequency. The magnitude of its spin is $\sqrt{2}\hbar$ and the component measured along its direction of motion, its helicity, must be $\pm\hbar$. These two possible helicities, called right-handed and left-handed, correspond to the two possible circular polarization states of the photon.

To illustrate the significance of these formulae, the annihilation of a particle with its antiparticle in free space must result in the creation of at least *two* photons for the following reason. In the center of momentum frame, the colliding antiparticles have no net momentum, whereas a single photon always has momentum (since, as we have seen, it is determined by the photon's frequency or wavelength, which cannot be zero). Hence, conservation of momentum (or equivalently, translational invariance) requires that at least two photons are created, with zero net momentum. (However, it is possible if the system interacts with another particle or field for the annihilation to produce one photon, as when a positron annihilates with a bound atomic electron, it is possible for only one photon to be emitted, as the nuclear Coulomb field breaks translational symmetry.) The energy of the two photons, or, equivalently, their frequency, may be determined from conservation of four-momentum. Seen another way, the photon can be considered as its own antiparticle. The reverse process, pair production, is the dominant mechanism by which high-energy photons such as gamma rays lose energy while passing through matter. That process is the reverse of "annihilation to one photon" allowed in the electric field of an atomic nucleus.

The classical formulae for the energy and momentum of electromagnetic radiation can be re-expressed in terms of photon events. For example, the pressure of electromagnetic radiation on an object derives from the transfer of photon momentum per unit time and unit area to that object, since pressure is force per unit area and force is the change in momentum per unit time.

Experimental Checks on Photon Mass

Current commonly accepted physical theories imply or assume the photon to be strictly massless. If the photon is not a strictly massless particle, it would not move at the exact speed of light, c in vacuum. Its speed would be lower and depend on its frequency. Relativity would be unaffected by this; the so-called speed of light, c, would then not be the actual speed at which light moves, but a constant of nature which is the maximum speed that any object could theoretically attain in space-time. Thus, it would still be the speed of space-time ripples (gravitational waves and gravitons), but it would not be the speed of photons.

If a photon did have non-zero mass, there would be other effects as well. Coulomb's law would be modified and the electromagnetic field would have an extra physical degree of freedom. These effects yield more sensitive experimental probes of the photon mass than the frequency dependence of the speed of light. If Coulomb's law is not exactly valid, then that would allow the presence of an electric field to exist within a hollow conductor when it is subjected to an external electric field. This thus allows one to test Coulomb's law to very high precision. A null result of such an experiment has set a limit of $m \lesssim 10^{-14}$ eV/c^2.

Sharper upper limits on the speed of light have been obtained in experiments designed to detect effects caused by the galactic vector potential. Although the galactic vector potential is very large because the galactic magnetic field exists on very great length scales, only the magnetic field would be observable if the photon is massless. In the case that the photon has mass, the mass term $\frac{1}{2}m^2 A_\mu A^\mu$ would affect the galactic plasma.

The fact that no such effects are seen implies an upper bound on the photon mass of $m < 3\times10^{-27}$ eV/c^2. The galactic vector potential can also be probed directly by measuring the torque exerted on a magnetized ring. Such methods were used to obtain the sharper upper limit of 10^{-18}eV/c^2 (the equivalent of 1.07×10^{-27} atomic mass units) given by the Particle Data Group.

These sharp limits from the non-observation of the effects caused by the galactic vector potential have been shown to be model dependent. If the photon mass is generated via the Higgs mechanism then the upper limit of $m \lesssim 10^{-14}$ eV/c^2 from the test of Coulomb's law is valid.

Photons inside superconductors do develop a nonzero effective rest mass; as a result, electromagnetic forces become short-range inside superconductors.

Historical Development

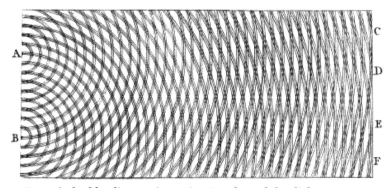

Thomas Young's double-slit experiment in 1801 showed that light can act as a wave,
helping to invalidate early particle theories of light.

In most theories up to the eighteenth century, light was pictured as being made up of particles. Since particle models cannot easily account for the refraction, diffraction and birefringence of light, wave theories of light were proposed by René Descartes (1637), Robert Hooke (1665), and Christiaan Huygens (1678); however, particle models remained dominant, chiefly due to the influence of Isaac Newton. In the early nineteenth century, Thomas Young and August Fresnel clearly demonstrated the interference and diffraction of light and by 1850 wave models were generally accepted. In 1865, James Clerk Maxwell's prediction that light was an electromagnetic wave—which was confirmed experimentally in 1888 by Heinrich Hertz's detection of radio waves—seemed to be the final blow to particle models of light.

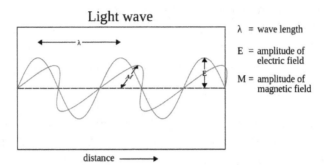

In 1900, Maxwell's theoretical model of light as oscillating electric and magnetic fields seemed complete. However, several observations could not be explained by any wave model of electromagnetic radiation, leading to the idea that light-energy was packaged into *quanta* described by E=hv. Later experiments showed that these light-quanta also carry momentum and, thus, can be considered particles: the *photon* concept was born, leading to a deeper understanding of the electric and magnetic fields themselves.

The Maxwell wave theory, however, does not account for *all* properties of light. The Maxwell theory predicts that the energy of a light wave depends only on its intensity, not on its frequency; nevertheless, several independent types of experiments show that the energy imparted by light to atoms depends only on the light's frequency, not on its intensity. For example, some chemical reactions are provoked only by light of frequency higher than a certain threshold; light of frequency lower than the threshold, no matter how intense, does not initiate the reaction. Similarly, electrons can be ejected from a metal plate by shining light of sufficiently high frequency on it (the photoelectric effect); the energy of the ejected electron is related only to the light's frequency, not to its intensity.

At the same time, investigations of blackbody radiation carried out over four decades (1860–1900) by various researchers culminated in Max Planck's hypothesis that the energy of *any* system that absorbs or emits electromagnetic radiation of frequency v is an integer multiple of an energy quantum $E = hv$. As shown by Albert Einstein, some form of energy quantization *must* be assumed to account for the thermal equilibrium observed between matter and electromagnetic radiation; for this explanation of the photoelectric effect, Einstein received the 1921 Nobel Prize in physics.

Since the Maxwell theory of light allows for all possible energies of electromagnetic radiation, most physicists assumed initially that the energy quantization resulted from some unknown constraint on the matter that absorbs or emits the radiation. In 1905, Einstein was the first to propose that energy quantization was a property of electromagnetic radiation itself. Although he accepted the validity of Maxwell's theory, Einstein pointed out that many anomalous experiments could be explained if the *energy* of a Maxwellian light wave were localized into point-like quanta that move independently of one another, even if the wave itself is spread continuously over space. In 1909 and 1916, Einstein showed that, if Planck's law of black-body radiation is accepted, the energy quanta must also carry momentum $p = h/\lambda$, making them full-fledged particles.

This photon momentum was observed experimentally by Arthur Compton, for which he received the Nobel Prize in 1927. The pivotal question was then: how to unify Maxwell's wave theory of light with its experimentally observed particle nature? The answer to this question occupied Albert Einstein for the rest of his life, and was solved in quan-tum electrodynamics and its successor, the Standard Model.

Einstein's Light Quantum

Unlike Planck, Einstein entertained the possibility that there might be actual physical quanta of light—what we now call photons. He noticed that a light quantum with energy proportional to its frequency would explain a number of troubling puzzles and paradoxes, including an unpublished law by Stokes, the ultraviolet catastrophe, and the photoelectric effect. Stokes's law said simply that the frequency of fluorescent light cannot be greater than the frequency of the light (usually ultraviolet) inducing it. Einstein eliminated the ultraviolet catastrophe by imagining a gas of photons behaving like a gas of electrons that he had previously considered. He was advised by a colleague to be careful how he wrote up this paper, in order to not challenge Planck, a powerful figure in physics, too directly, and indeed the warning was justified, as Planck never forgave him for writing it.

Early Objections

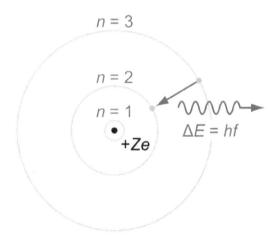

Up to 1923, most physicists were reluctant to accept that light itself was quantized. Instead, they tried to explain photon behavior by quantizing only *matter*, as in the Bohr model of the hydrogen atom (shown here). Even though these semiclassical models were only a first approximation, they were accurate for simple systems and they led to quantum mechanics.

Einstein's 1905 predictions were verified experimentally in several ways in the first two decades of the 20th century, as recounted in Robert Millikan's Nobel lecture. However, before Compton's experiment showed that photons carried momentum proportional to their wave number (1922), most physicists were reluctant to believe that electromagnetic radiation itself might be particulate. (for example, the Nobel lectures of Wien, Planck and Millikan.) Instead, there was a widespread belief that energy quantization resulted from some unknown constraint on the matter that absorbed or emitted radiation. Attitudes changed over time. In part, the change can be traced to experiments such as Compton scattering, where it was much more difficult not to ascribe quantization to light itself to explain the observed results.

Even after Compton's experiment, Niels Bohr, Hendrik Kramers and John Slater made one last attempt to preserve the Maxwellian continuous electromagnetic field model of light, the so-called BKS model. To account for the data then available, two drastic hypotheses had to be made:

1. Energy and momentum are conserved only on the average in interactions between matter and radiation, but not in elementary processes such as absorption and emission. This allows one to reconcile the discontinuously changing energy of the atom (the jump between energy states) with the continuous release of energy as radiation.

2. Causality is abandoned. For example, spontaneous emissions are merely emissions stimulated by a "virtual" electromagnetic field.

However, refined Compton experiments showed that energy–momentum is conserved extraordinarily well in elementary processes; and also that the jolting of the electron and the generation of a new photon in Compton scattering obey causality to within 10 ps. Accordingly, Bohr and his co-workers gave their model "as honorable a funeral as possible". Nevertheless, the failures of the BKS model inspired Werner Heisenberg in his development of matrix mechanics.

A few physicists persisted in developing semiclassical models in which electromagnetic radiation is not quantized, but matter appears to obey the laws of quantum mechanics. Although the evidence from chemical and physical experiments for the existence of photons was overwhelming by the 1970s, this evidence could not be considered as *absolutely* definitive; since it relied on the interaction of light with matter, and a sufficiently complete theory of matter could in principle account for the evidence. Nevertheless, *all* semiclassical theories were refuted definitively in the 1970s and 1980s by photon-correlation experiments. Hence, Einstein's hypothesis that quantization is a property of light itself is considered to be proven.

Wave–particle Duality and Uncertainty Principles

Photons, like all quantum objects, exhibit wave-like and particle-like properties. Their dual wave–particle nature can be difficult to visualize. The photon displays clearly wave-like phenomena such as diffraction and interference on the length scale of its wavelength. For example, a single photon passing through a double-slit experiment exhibits interference phenomena but only if no measure was made at the slit. A single photon passing through a double-slit experiment lands on the screen with a probability distribution given by its interference pattern determined by Maxwell's equations.

However, experiments confirm that the photon is *not* a short pulse of electromagnetic radiation; it does not spread out as it propagates, nor does it divide when it encounters a beam splitter. Rather, the photon seems to be a point-like particle since it is absorbed or emitted *as a whole* by arbitrarily small systems, systems much smaller than its wavelength, such as an atomic nucleus ($\approx 10^{-15}$ m across) or even the point-like electron. Nevertheless, the photon is *not* a point-like particle whose trajectory is shaped probabilistically by the electromagnetic field, as conceived by Einstein and others; that hypothesis was also refuted by the photon-correlation experiments cited above. According to our present understanding, the electromagnetic field itself is produced by photons, which in turn result from a local gauge symmetry and the laws of quantum field theory.

Photons in a Mach–Zehnder interferometer exhibit wave-like interference and particle-like detection at single-photon detectors.

A key element of quantum mechanics is Heisenberg's uncertainty principle, which forbids the simultaneous measurement of the position and momentum of a particle along the same direction. Remarkably, the uncertainty principle for charged, material particles *requires* the quantization of light into photons, and even the frequency dependence of the photon's energy and momentum.

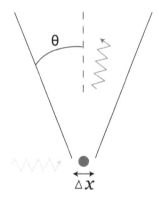

Heisenberg's thought experiment for locating an electron (shown in blue) with a high-resolution gamma-ray microscope. The incoming gamma ray (shown in green) is scattered by the electron up into the microscope's aperture angle θ. The scattered gamma ray is shown in red. Classical optics shows that the electron position can be resolved only up to an uncertainty Δx that depends on θ and the wavelength λ of the incoming light.

An elegant illustration of the uncertainty principle is Heisenberg's thought experiment for locating an electron with an ideal microscope. The position of the electron can be determined to within the resolving power of the microscope, which is given by a formula from classical optics

$$\Delta x \sim \frac{\lambda}{\sin\theta}$$

where θ is the aperture angle of the microscope and λ is the wavelength of the light used to observe the electron. Thus, the position uncertainty Δx can be made arbitrarily small by reducing the wavelength λ. Even if the momentum of the electron is initially known, the light impinging on the electron will give it a momentum "kick" Δp of some unknown amount, rendering the momentum of the electron uncertain. If light were *not* quantized into photons, the uncertainty Δp could be made arbitrarily small by reducing the light's intensity. In that case, since the wavelength and intensity of light can be varied independently, one could simultaneously determine the position and momentum to arbitrarily high accuracy, violating the uncertainty principle. By contrast, Einstein's formula for photon momentum preserves the uncertainty principle; since the photon is scattered anywhere within the aperture, the uncertainty of momentum transferred equals

$$\Delta p \sim p_{\text{photon}} \sin\theta = \frac{h}{\lambda}\sin\theta$$

giving the product $\Delta x \Delta p \sim h,$, which is Heisenberg's uncertainty principle. Thus, the entire world is quantized; both matter and fields must obey a consistent set of quantum laws, if either one is to be quantized.

The analogous uncertainty principle for photons forbids the simultaneous measurement of the number n of photons in an electromagnetic wave and the phase ϕ of that wave

$$Än Ä\phi > 1$$

Both photons and electrons create analogous interference patterns when passed through a double-slit experiment. For photons, this corresponds to the interference of a Maxwell light wave whereas, for material particles (electron), this corresponds to the interference of the Schrödinger wave equation. Although this similarity might suggest that Maxwell's equations describing the photon's electromagnetic wave are simply Schrödinger's equation for photons, most physicists do not agree. For one thing, they are mathematically different; most obviously, Schrödinger's one equation for the electron solves for a complex field, whereas Maxwell's four equations solve for real fields. More generally, the normal concept of a Schrödinger probability wave function cannot be applied to photons. As photons are massless, they cannot be localized without being destroyed; technically, photons cannot have a position eigenstate $|r\rangle$,, and, thus, the normal Heisenberg uncertainty principle $\Delta x \Delta p > h/2$ does not pertain to photons. A few substitute wave functions have been suggested for the photon, but they have not come into general use. Instead, physicists generally accept the second-quantized theory of photons described below, quantum electrodynamics, in which photons are quantized excitations of electromagnetic modes.

Another interpretation, that avoids duality, is the De Broglie–Bohm theory: known also as the *pilot-wave model*. In that theory, the photon is both, wave and particle. *"This idea seems to me so natural and simple, to resolve the wave-particle dilemma in such a clear and ordinary way, that it is a great mystery to me that it was so generally ignored"*, J.S.Bell.

Bose–Einstein Model of a Photon Gas

In 1924, Satyendra Nath Bose derived Planck's law of black-body radiation without using any electromagnetism, but rather by using a modification of coarse-grained counting of phase space. Einstein showed that this modification is equivalent to assuming that photons are rigorously identical and that it implied a "mysterious non-local interaction", now understood as the requirement for a symmetric quantum mechanical state. This work led to the concept of coherent states and the development of the laser. In the same papers, Einstein extended Bose's formalism to material particles (bosons) and predicted that they would condense into their lowest quantum state at low enough temperatures; this Bose–Einstein condensation was observed experimentally in 1995. It was later used by Lene Hau to slow, and then completely stop, light in 1999 and 2001.

The modern view on this is that photons are, by virtue of their integer spin, bosons (as opposed to fermions with half-integer spin). By the spin-statistics theorem, all bosons obey Bose–Einstein statistics (whereas all fermions obey Fermi–Dirac statistics).

Stimulated and Spontaneous Emission

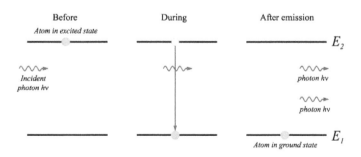

Stimulated emission (in which photons "clone" themselves) was predicted by Einstein in his kinetic analysis, and led to the development of the laser. Einstein's derivation inspired further developments in the quantum treatment of light, which led to the statistical interpretation of quantum mechanics.

In 1916, Einstein showed that Planck's radiation law could be derived from a semi-classical, statistical treatment of photons and atoms, which implies a link between the rates at which atoms emit and absorb photons. The condition follows from the assumption that functions of the emission and absorption of radiation by the atoms are independent of each other, and that thermal equilibrium is made by way of the radiation's interaction with the atoms. Consider a cavity in thermal equilibrium with all parts of itself and filled with electromagnetic radiation and that the atoms can emit and absorb that radiation. Thermal equilibrium requires that the energy density $\rho(\nu)$ of photons with frequency ν (which is proportional to their number density) is, on average, constant in time; hence, the rate at which photons of any particular frequency are *emitted* must equal the rate at which they *absorb* them.

Einstein began by postulating simple proportionality relations for the different reaction rates involved. In his model, the rate R_{ji} for a system to *absorb* a photon of frequency ν and transition from a lower energy E_j to a higher energy E_i is proportional to the number N_j of atoms with energy and to the energy density $\rho(\nu)$ of ambient photons of that frequency,

$$R_{ji} = N_j B_{ji} \rho(\nu)$$

where B_{ji} is the rate constant for absorption. For the reverse process, there are two possibilities: spontaneous emission of a photon, or the emission of a photon initiated by the interaction of the atom with a passing photon and the return of the atom to the lower-energy state. Following Einstein's approach, the corresponding rate R_{ij} for the emission of photons of frequency ν and transition from a higher energy E_i to a lower energy E_j is

$$R_{ij} = N_i A_{ij} + N_i B_{ij} \rho(\nu)$$

where A_{ij} is the rate constant for emitting a photon spontaneously, and B_{ij} is the rate constant for emissions in response to ambient photons (induced or stimulated emission). In thermodynamic equilibrium, the number of atoms in state i and those in state j must, on average, be constant; hence, the rates R_{ji} and R_{ij} must be equal. Also, by arguments analogous to the derivation of Boltzmann statistics, the ratio of N_i and N_j is $g_i / g_j \exp(E_j - E_i)/(kT)$, where $g_{i,j}$ are the degeneracy of the state i and that of j, respectively, $E_{i,j}$ their energies, k the Boltzmann constant and T the system's temperature. From this, it is readily derived that $g_i B_{ij} = g_j B_{ji}$ and

$$A_{ij} = \frac{8\pi h \nu^3}{c^3} B_{ij}.$$

The A and Bs are collectively known as the *Einstein coefficients*.

Einstein could not fully justify his rate equations, but claimed that it should be possible to calculate the coefficients A_{ij}, B_{ji} and B_{ij} once physicists had obtained "mechanics and electrodynamics modified to accommodate the quantum hypothesis". In fact, in 1926, Paul Dirac derived the B_{ij} rate constants by using a semiclassical approach, and, in 1927, succeeded in deriving *all* the rate constants from first principles within the framework of quantum theory. Dirac's work was the foundation of quantum electrodynamics, i.e., the quantization of the electromagnetic field itself. Dirac's approach is also called *second quantization* or quantum field theory; earlier quantum mechanical treatments only treat material particles as quantum mechanical, not the electromagnetic field.

Einstein was troubled by the fact that his theory seemed incomplete, since it did not determine the *direction* of a spontaneously emitted photon. A probabilistic nature of light-particle motion was first considered by Newton in his treatment of birefringence and, more generally, of the splitting of light beams at interfaces into a transmitted beam and a reflected beam. Newton hypothesized that hidden variables in the light particle determined which of the two paths a single photon would take. Similarly, Einstein hoped for a more complete theory that would leave nothing to chance, beginning his separation from quantum mechanics. Ironically, Max Born's probabilistic interpretation of the wave function was inspired by Einstein's later work searching for a more complete theory.

Second Quantization and High Energy Photon Interactions

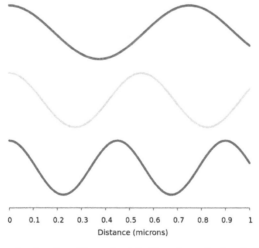

Different *electromagnetic modes* (such as those depicted here) can be treated as independent simple harmonic oscillators. A photon corresponds to a unit of energy E=hv in its electromagnetic mode.

In 1910, Peter Debye derived Planck's law of black-body radiation from a relatively simple assumption. He correctly decomposed the electromagnetic field in a cavity into its Fourier modes, and assumed that the energy in any mode was an integer multiple of hv, where v is the frequency of the electromagnetic mode. Planck's law of black-body radiation follows immediately as a geometric sum. However, Debye's approach failed to give the correct formula for the energy fluctuations of blackbody radiation, which were derived by Einstein in 1909.

In 1925, Born, Heisenberg and Jordan reinterpreted Debye's concept in a key way. As may be shown classically, the Fourier modes of the electromagnetic field—a complete set of electromagnetic plane waves indexed by their wave vector k and polarization state—are equivalent to a set of uncoupled simple harmonic oscillators. Treated quantum mechanically, the energy levels of such oscillators are known to be $E = nh\nu$, where ν is the oscillator frequency. The key new step was to identify an electromagnetic mode with energy $E = nh\nu$ as a state with n photons, each of energy $h\nu$. This approach gives the correct energy fluctuation formula.

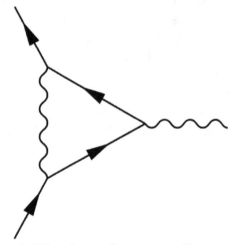

In quantum field theory, the probability of an event is computed by summing the probability amplitude (a complex number) for all possible ways in which the event can occur, as in the Feynman diagram shown here; the probability equals the square of the modulus of the total amplitude.

Dirac took this one step further. He treated the interaction between a charge and an electromagnetic field as a small perturbation that induces transitions in the photon states, changing the numbers of photons in the modes, while conserving energy and momentum overall. Dirac was able to derive Einstein's A_{ij} and B_{ij} coefficients from first principles, and showed that the Bose–Einstein statistics of photons is a natural consequence of quantizing the electromagnetic field correctly (Bose's reasoning went in the opposite direction; he derived Planck's law of black-body radiation by *assuming* B–E statistics). In Dirac's time, it was not yet known that all bosons, including photons, must obey Bose–Einstein statistics.

Dirac's second-order perturbation theory can involve virtual photons, transient intermediate states of the electromagnetic field; the static electric and magnetic interactions are mediated by such virtual photons. In such quantum field theories, the probability amplitude of observable events is calculated by summing over *all* possible intermediate steps, even ones that are unphysical; hence, virtual photons are not constrained to satisfy

$E = pc$, and may have extra polarization states; depending on the gauge used, virtual photons may have three or four polarization states, instead of the two states of real photons. Although these transient virtual photons can never be observed, they contribute measurably to the probabilities of observable events. Indeed, such second-order and higher-order perturbation calculations can give apparently infinite contributions to the sum. Such unphysical results are corrected for using the technique of renormalization.

Other virtual particles may contribute to the summation as well; for example, two photons may interact indirectly through virtual electron–positron pairs. In fact, such photon-photon scattering, as well as electron-photon scattering, is meant to be one of the modes of operations of the planned particle accelerator, the International Linear Collider.

In modern physics notation, the quantum state of the electromagnetic field is written as a Fock state, a tensor product of the states for each electromagnetic mode

$$|n_{k_0}\rangle \otimes |n_{k_1}\rangle \otimes \dots \otimes |n_{k_n}\rangle \dots$$

where $|n_{k_i}\rangle$ represents the state in which n_{k_i} photons are in the mode k_i. In this notation, the creation of a new photon in mode k_i (e.g., emitted from an atomic transition) is written as $|n_{k_i}\rangle \rightarrow |n_{k_i}+1\rangle$. This notation merely expresses the concept of Born, Heisenberg and Jordan described above, and does not add any physics.

The Hadronic Properties of the Photon

Measurements of the interaction between energetic photons and hadrons show that the interaction is much more intense than expected by the interaction of merely photons with the hadron's electric charge. Furthermore, the interaction of energetic photons with protons is similar to the interaction of photons with neutrons in spite of the fact that the electric charge structures of protons and neutrons are substantially different. A theory called Vector Meson Dominance (VMD) was developed to explain this effect. According to VMD, the photon is a superposition of the pure electromagnetic photon which interacts only with electric charges and vector meson. However, if experimentally probed at very short distances, the intrinsic structure of the photon is recognized as a flux of quark and gluon components, quasi-free according to asymptotic freedom in QCD and described by the photon structure function. A comprehensive comparison of data with theoretical predictions is presented in a recent review.

The Photon as a Gauge Boson

The electromagnetic field can be understood as a gauge field, i.e., as a field that results from requiring that a gauge symmetry holds independently at every position in space-time. For the electromagnetic field, this gauge symmetry is the Abelian U(1) symmetry of complex numbers of absolute value 1, which reflects the ability to vary the phase of a complex field without affecting observables or real valued functions made from it, such as the energy or the Lagrangian.

The quanta of an Abelian gauge field must be massless, uncharged bosons, as long as the symmetry is not broken; hence, the photon is predicted to be massless, and to have zero electric charge and integer spin. The particular form of the electromagnetic interaction specifies that the photon must have spin ± 1; thus, its helicity must be $\pm \hbar$. . These two spin components correspond to the classical concepts of right-handed and left-handed circularly polarized light. However, the transient virtual photons of quantum electrodynamics may also adopt unphysical polarization states.

In the prevailing Standard Model of physics, the photon is one of four gauge bosons in the electroweak interaction; the other three are denoted W^+, W^- and Z^0 and are responsible for the weak interaction. Unlike the photon, these gauge bosons have mass, owing to a mechanism that breaks their SU(2) gauge symmetry. The unification of the photon with W and Z gauge bosons in the electroweak interaction was accomplished by Sheldon Glashow, Abdus Salam and Steven Weinberg, for which they were awarded the 1979 Nobel Prize in physics. Physicists continue to hypothesize grand unified theories that connect these four gauge bosons with the eight gluon gauge bosons of quantum chromodynamics; however, key predictions of these theories, such as proton decay, have not been observed experimentally.

Contributions to the Mass of a System

The energy of a system that emits a photon is *decreased* by the energy E of the photon as measured in the rest frame of the emitting system, which may result in a reduction in mass in the amount E/c^2. Similarly, the mass of a system that absorbs a photon is *increased* by a corresponding amount. As an application, the energy balance of nuclear reactions involving photons is commonly written in terms of the masses of the nuclei involved, and terms of the form E/c^2 for the gamma photons (and for other relevant energies, such as the recoil energy of nuclei).

This concept is applied in key predictions of quantum electrodynamics. In that theory, the mass of electrons (or, more generally, leptons) is modi-fied by including the mass contributions of virtual photons, in a technique known as renormalization. Such "radiative corrections" contribute to a number of predictions of QED, such as the magnetic dipole moment of leptons, the Lamb shift, and the hyperfine structure of bound lepton pairs, such as muonium and positronium.

Since photons contribute to the stress–energy tensor, they exert a gravitational attraction on other objects, according to the theory of general relativity. Conversely, photons are themselves affected by gravity; their normally straight trajectories may be bent by warped spacetime, as in gravitational lensing, and their frequencies may be lowered by moving to a higher gravitational potential, as in the Pound–Rebka experiment. However, these effects are not specific to photons; exactly the same effects would be predicted for classical electromagnetic waves.

Photons in Matter

Light that travels through transparent matter does so at a lower speed than c, the speed of light in a vacuum. For example, photons engage in so many collisions on the way from the core of the sun that radiant energy can take about a million years to reach the surface; however, once in open space, a photon takes only 8.3 minutes to reach Earth. The factor by which the speed is decreased is called the refractive index of the material. In a classical wave picture, the slowing can be explained by the light inducing electric polarization in the matter, the polarized matter radiating new light, and that new light interfering with the original light wave to form a delayed wave. In a particle picture, the slowing can instead be described as a blending of the photon with quantum excitations of the matter to produce quasi-particles known as polariton (other quasi-particles are phonons and excitons); this polariton has a nonzero effective mass, which means that it cannot travel at c. Light of different frequencies may travel through matter at different speeds; this is called dispersion. In some cases, it can result in extremely slow speeds of light in matter. The effects of photon interactions with other quasi-particles may be observed directly in Raman scattering and Brillouin scattering.

Photons can also be absorbed by nuclei, atoms or molecules, provoking transitions between their energy levels. A classic example is the molecular transition of retinal ($C_{20}H_{28}O$), which is responsible for vision, as discovered in 1958 by Nobel laureate biochemist George Wald and co-workers. The absorption provokes a cis-trans isomerization that, in combination with other such transitions, is transduced into nerve impulses. The absorption of photons can even break chemical bonds, as in the photodissociation of chlorine; this is the subject of photochemistry.

Technological Applications

Photons have many applications in technology. These examples are chosen to illustrate applications of photons *per se*, rather than general optical devices such as lenses, etc. that could operate under a classical theory of light. The laser is an extremely important application and is discussed above under stimulated emission.

Individual photons can be detected by several methods. The classic photomultiplier tube exploits the photoelectric effect: a photon of sufficient energy strikes a metal plate and knocks free an electron, initiating an ever-amplifying avalanche of electrons. Semiconductor charge-coupled device chips use a similar effect: an incident photon generates a charge on a microscopic capacitor that can be detected. Other detectors such as Geiger counters use the ability of photons to ionize gas molecules contained in the device, causing a detectable change of conductivity of the gas.

Planck's energy formula $E = h\nu$ is often used by engineers and chemists in design, both to compute the change in energy resulting from a photon absorption and to determine the frequency of the light emitted from a given photon emission. For example, the emission spectrum of a gas-discharge lamp can be altered by filling it with (mixtures of) gases with different electronic energy level configurations.

Under some conditions, an energy transition can be excited by "two" photons that individually would be insufficient. This allows for higher resolution microscopy, because the sample absorbs energy only in the spectrum where two beams of different colors overlap significantly, which can be made much smaller than the excitation volume of a single beam. Moreover, these photons cause less damage to the sample, since they are of lower energy.

In some cases, two energy transitions can be coupled so that, as one system absorbs a photon, another nearby system "steals" its energy and re-emits a photon of a different frequency. This is the basis of fluorescence resonance energy transfer, a technique that is used in molecular biology to study the interaction of suitable proteins.

Several different kinds of hardware random number generators involve the detection of single photons. In one example, for each bit in the random sequence that is to be produced, a photon is sent to a beam-splitter. In such a situation, there are two possible outcomes of equal probability. The actual outcome is used to determine whether the next bit in the sequence is "0" or "1".

Recent Research

Much research has been devoted to applications of photons in the field of quantum optics. Photons seem well-suited to be elements of an extremely fast quantum computer, and the quantum entanglement of photons is a focus of research. Nonlinear optical processes are another active research area, with topics such as two-photon absorption, self-phase modulation, modulational instability and optical parametric oscillators. However, such processes generally do not require the assumption of photons *per se*; they may often be modeled by treating atoms as nonlinear oscillators. The nonlinear process of spontaneous parametric down conversion is often used to produce single-photon states. Finally, photons are essential in some aspects of optical communication, especially for quantum cryptography.

Lepton

A lepton is an elementary, half-integer spin (spin $\frac{1}{2}$) particle that does not undergo strong interactions. Two main classes of leptons exist: charged leptons (also known as the *electron-like* leptons), and neutral leptons (better known as neutrinos). Charged leptons can combine with other particles to form various composite particles such as atoms and positronium, while neutrinos rarely interact with anything, and are consequently rarely observed. The best known of all leptons is the electron.

There are six types of leptons, known as *flavours*, forming three *generations*. The first generation is the *electronic leptons*, comprising the electron (e−) and electron neutrino (v e); the second is the *muonic leptons*, comprising the muon (μ−) and muon neutrino (v μ); and the third is the *tauonic leptons*, comprising the tau (τ−) and the tau neutrino (v τ). Electrons have the least mass of all the charged leptons. The heavier muons and taus will rapidly change into electrons and neutrinos through a process of particle decay: the transformation from a higher mass state to a lower mass state. Thus electrons are stable and the most common charged lepton in the universe, whereas muons and taus can only be produced in high energy collisions (such as those involving cosmic rays and those carried out in particle accelerators).

Leptons have various intrinsic properties, including electric charge, spin, and mass. Unlike quarks however, leptons are not subject to the strong interaction, but they are subject to the other three fundamental interactions: gravitation, electromagnetism (excluding neutrinos, which are electrically neutral), and the weak interaction.

For every lepton flavor there is a corresponding type of antiparticle, known as an antilepton, that differs from the lepton only in that some of its properties have equal magnitude but opposite sign. However, according to certain theories, neutrinos may be their own antiparticle, but it is not currently known whether this is the case or not.

The first charged lepton, the electron, was theorized in the mid-19th century by several scientists and was discovered in 1897 by J. J. Thomson. The next lepton to be observed was the muon, discovered by Carl D. Anderson in 1936, which was classified as a meson at the time. After investigation, it was realized that the muon did not have the expected properties of a meson, but rather behaved like an electron, only with higher mass. It took until 1947 for the concept of "leptons" as a family of particle to be proposed. The first neutrino, the electron neutrino, was proposed by Wolfgang Pauli in 1930 to explain certain characteristics of beta decay. It was first observed in the Cowan–Reines neutrino experiment conducted by Clyde Cowan and Frederick Reines in 1956. The muon neutrino was discovered in 1962 by Leon M. Lederman, Melvin Schwartz and Jack Steinberger, and the tau discovered between 1974 and 1977 by Martin Lewis Perl and his colleagues from the Stanford Linear Accelerator Center and Lawrence Berkeley National Laboratory. The tau neutrino remained elusive until July 2000, when the DONUT collaboration from Fermilab announced its discovery.

Leptons are an important part of the Standard Model. Electrons are one of the components of atoms, alongside protons and neutrons. Exotic atoms with muons and taus instead of electrons can also be synthesized, as well as lepton–antilepton particles such as positronium.

Etymology

The name *lepton* comes from the Greek λεπτός *leptós*, "fine, small, thin" (neuter nominative/accusative singular form: λεπτόν *leptón*); the earliest attested form of the word is the Mycenaean Greek, *re-po-to*, written in Linear B syllabic script. *Lepton* was first used by physicist Léon Rosenfeld in 1948:

Following a suggestion of Prof. C. Møller, I adopt — as a pendant to "nucleon" — the denomination "lepton" (from λεπτός, small, thin, delicate) to denote a particle of small mass.

The etymology incorrectly implies that all the leptons are of small mass. When Rosenfeld named them, the only known leptons were electrons and muons, which are in fact of small mass — the mass of an electron (0.511 MeV/c^2) and the mass of a muon (with a value of 105.7 MeV/c^2) are fractions of the mass of the "heavy" proton (938.3 MeV/c^2). However, the mass of the tau (discovered in the mid 1970s) (1777 MeV/c^2) is nearly twice that of the proton, and about 3,500 times that of the electron.

History

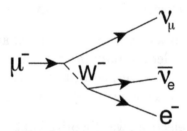

A muon transmutes into a muon neutrino by emitting a W− boson. The W− boson subsequently decays into an electron and an electron antineutrino.

Lepton nomenclature	
Particle name	**Antiparticle name**
Electron	Antielectron Positron
Electron neutrino	Electron antineutrino
Muon Mu lepton Mu	Antimuon Antimu lepton Antimu
Muon neutrino Muonic neutrino Mu neutrino	Muon antineutrino Muonic antineutrino Mu antineutrino
Tauon Tau lepton Tau	Antitauon Antitau lepton Antitau
Tauon neutrino Tauonic neutrino Tau neutrino	Tauon antineutrino Tauonic antineutrino Tau antineutrino

The first lepton identified was the electron, discovered by J.J. Thomson and his team of British physicists in 1897. Then in 1930 Wolfgang Pauli postulated the electron neutrino to preserve conservation of energy, conservation of momentum, and conservation of angular momentum in beta decay. Pauli theorized that an undetected particle was carrying away the difference between the energy, momentum, and angular momentum of the initial and observed final particles. The electron neutrino was simply called the neutrino, as it was not yet known that neutrinos came in different flavours (or different "generations").

Nearly 40 years after the discovery of the electron, the muon was discovered by Carl D. Anderson in 1936. Due to its mass, it was initially categorized as a meson rather than a lepton. It later became clear that the muon was much more similar to the electron than to mesons, as muons do not undergo the strong interaction, and thus the muon was reclassified: electrons, muons, and the (electron) neutrino were grouped into a new group of particles – the leptons. In 1962 Leon M. Lederman, Melvin Schwartz and Jack Steinberger showed that more than one type of neutrino exists by first detecting interactions of the muon neutrino, which earned them the 1988 Nobel Prize, although by then the different flavours of neutrino had already been theorized.

The tau was first detected in a series of experiments between 1974 and 1977 by Martin Lewis Perl with his colleagues at the SLAC LBL group. Like the electron and the muon, it too was expected to have an associated neutrino. The first evidence for tau neutrinos came from the observation of "missing" energy and momentum in tau decay, analogous to the "missing" energy and momentum in beta decay leading to the discovery of the electron neutrino. The first detection of tau neutrino interactions was announced in 2000 by the DONUT collaboration at Fermilab, making it the latest particle of the Standard Model to have been directly observed, apart from the Higgs boson, which probably has been discovered in 2012.

Although all present data is consistent with three generations of leptons, some particle physicists are searching for a fourth generation. The current lower limit on the mass of such a fourth charged lepton is 100.8 GeV/c^2, while its associated neutrino would have a mass of at least 45.0 GeV/c^2.

Properties

Spin and Chirality

Left-handed and right-handed helicities

Leptons are spin-$\frac{1}{2}$ particles. The spin-statistics theorem thus implies that they are fermions and thus that they are subject to the Pauli exclusion principle; no two leptons of the same species can be in exactly the same state at the same time. Furthermore, it means that a lepton can have only two possible spin states, namely up or down.

A closely related property is chirality, which in turn is closely related to a more easily visualized property called helicity. The helicity of a particle is the direction of its spin relative to its momentum; particles with spin in the same direction as their momentum are called *right-handed* and otherwise they are called *left-handed*. When a particle is mass-less, the direction of its momentum relative to its spin is frame independent, while for massive particles it is possible to 'overtake' the particle by a Lorentz transformation flipping the helicity. Chirality is a technical property (defined through the transformation behaviour under the Poincaré group) that agrees with helicity for (approximately) massless particles and is still well defined for massive particles.

In many quantum field theories—such as quantum electrodynamics and quantum chromodynamics—left and right-handed fermions are identical. However, in the Standard Model left-handed and right-handed fermions are treated asymmetrically. Only left-handed fermions participate in the weak interaction, while there are no right-handed neutrinos. This is an example of parity violation. In the literature left-handed fields are often denoted by a capital L subscript (e.g. e$-_L$) and right-handed fields are denoted by a capital R subscript.

Electromagnetic Interaction

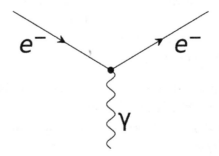

Lepton–photon interaction

One of the most prominent properties of leptons is their electric charge, Q. The electric charge determines the strength of their electromagnetic interactions. It determines the strength of the electric field generated by the particle and how strong-ly the particle reacts to an external electric or magnetic field. Each generation contains one lepton with $Q = -e$ (conventionally the charge of a particle is expressed in units of the elementary charge) and one lepton with zero electric charge. The lepton with electric charge is commonly simply referred to as a 'charged lepton' while the neutral lepton is called a neutrino. For example, the first generation consists of the electron e– with a negative electric charge and the electrically neutral electron neutrino ve.

In the language of quantum field theory the electromagnetic interaction of the charged leptons is expressed by the fact that the particles interact with the quantum of the electromagnetic field, the photon. The Feynman diagram of the electron-photon interaction is shown on the right.

Because leptons possess an intrinsic rotation in the form of their spin, charged leptons generate a magnetic field. The size of their magnetic dipole moment μ is given by,

$$\mu = g\frac{Q\hbar}{4m},$$

where m is the mass of the lepton and g is the so-called g-factor for the lepton. First order approximation quantum mechanics predicts that the g-factor is 2 for all leptons. However, higher order quantum effects caused by loops in Feynman diagrams introduce corrections to this value. These corrections, referred to as the anomalous magnetic dipole moment, are very sensitive to the details of a quantum field theory model and thus provide the opportunity for precision tests of the standard model. The theoretical and measured values for the electron anomalous magnetic dipole moment are within agreement within eight significant figures.

Weak Interaction

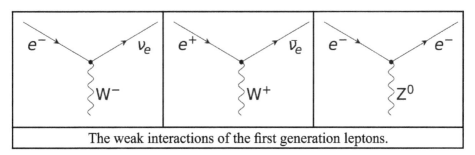

The weak interactions of the first generation leptons.

In the Standard Model, the left-handed charged lepton and the left-handed neutrino are arranged in doublet $(ve_L, e-_L)$ that transforms in the spinor representation $(T = \frac{1}{2})$ of the weak isospin SU(2) gauge symmetry. This means that these particles are eigenstates of the isospin projection T_3 with eigenvalues $\frac{1}{2}$ and $-\frac{1}{2}$ respectively. In the meantime, the right-handed charged lepton transforms as a weak isospin scalar $(T = 0)$ and thus does not participate in the weak interaction, while there is no right-handed neutrino at all.

The Higgs mechanism recombines the gauge fields of the weak isospin SU(2) and the weak hypercharge U(1) symmetries to three massive vector bosons (W+, W−, Zo) mediating the weak interaction, and one massless vector boson, the photon, responsible for the electromagnetic interaction. The electric charge Q can be calculated from the isospin projection T_3 and weak hypercharge Y_W through the Gell-Mann–Nishijima formula,

$$Q = T_3 + Y_W/2$$

To recover the observed electric charges for all particles the left-handed weak isospin doublet $(\nu e_L, e-_L)$ must thus have $Y_W = -1$, while the right-handed isospin scalar e–R must have $Y_W = -2$. The interaction of the leptons with the massive weak interaction vector bosons is shown in the figure on the left.

Mass

In the Standard Model each lepton starts out with no intrinsic mass. The charged leptons (i.e. the electron, muon, and tau) obtain an effective mass through interaction with the Higgs field, but the neutrinos remain massless. For technical reasons the masslessness of the neutrinos implies that there is no mixing of the different generations of charged leptons as there is for quarks. This is in close agreement with current experimental observations.

However, it is known from experiments – most prominently from observed neutrino oscillations – that neutrinos do in fact have some very small mass, probably less than 2 eV/c^2. This implies the existence of physics beyond the Standard Model. The currently most favoured extension is the so-called seesaw mechanism, which would explain both why the left-handed neutrinos are so light compared to the corresponding charged leptons, and why we have not yet seen any right-handed neutrinos.

Leptonic Numbers

The members of each generation's weak isospin doublet are assigned leptonic numbers that are conserved under the Standard Model. Electrons and electron neutrinos have an *electronic number* of $L_e = 1$, while muons and muon neutrinos have a *muonic number* of $L_\mu = 1$, while tau particles and tau neutrinos have a *tauonic number* of $L_\tau = 1$. The antileptons have their respective generation›s leptonic numbers of –1.

Conservation of the leptonic numbers means that the number of leptons of the same type remains the same, when particles interact. This implies that leptons and antileptons must be created in pairs of a single generation. For example, the following processes are allowed under conservation of leptonic numbers:

$$\begin{pmatrix} \nu_e \\ e^- \end{pmatrix}, \begin{pmatrix} \nu_\mu \\ \mu^- \end{pmatrix}, \begin{pmatrix} \nu_\tau \\ \tau^- \end{pmatrix}$$

Each generation forms a weak isospin doublet.

$\quad$ e– + e+ → γ + γ,

$\quad$ τ– + τ+ → Z0 + Z0,

but not these:

$\quad$ γ → e– + μ+, W– → e– + ντ, Z0 → μ– + τ+.

However, neutrino oscillations are known to violate the conservation of the individual leptonic numbers. Such a violation is considered to be smoking gun evidence for physics beyond the Standard Model. A much stronger conservation law is the conservation of the total number of leptons (L), conserved even in the case of neutrino oscillations, but even it is still violated by a tiny amount by the chiral anomaly.

Universality

The coupling of the leptons to gauge bosons are flavour-independent (i.e., the interactions between leptons and gauge bosons are the same for all leptons). This property is called *lepton universality* and has been tested in measurements of the tau and muon lifetimes and of Z boson partial decay widths, particularly at the Stanford Linear Collider (SLC) and Large Electron-Positron Collider (LEP) experiments.

The decay rate (Γ) of muons through the process $\mu- \rightarrow e- + ve + \nu\mu$ is approximately given by an expression of the form.

$$\Gamma\left(\mu^- \rightarrow e^- +\overline{v}_e +v_\mu\right)= K_1 G_F^2 m_\mu^5,$$

where K_1 is some constant, and G_F is the Fermi coupling constant. The decay rate of tau particles through the process $\tau- \rightarrow e- + ve + v\tau$ is given by an expression of the same form

$$\Gamma\left(\tau^- \rightarrow e^- +\overline{v}_e +v_\tau\right)= K_2 G_F^2 m_\tau^5,$$

where K_2 is some constant. Muon–Tauon universality implies that $K_1 = K_2$. On the other hand, electron–muon universality implies

$$\Gamma\left(\tau^- \rightarrow e^- +\overline{v}_e +v_\tau\right)=\Gamma\left(\tau^- \rightarrow \mu^- +\overline{v}_\mu +v_\tau\right).$$

This explains why the branching ratios for the electronic mode (17.85%) and muonic (17.36%) mode of tau decay are equal (within error).

Universality also accounts for the ratio of muon and tau lifetimes. The lifetime of a lepton (τ_l) is related to the decay rate by

$$\tau_l = \frac{B\left(l^- \rightarrow e^- +\overline{v}_e +v_l\right)}{\Gamma\left(l^- \rightarrow e^- +\overline{v}_e +v_l\right)},$$

where $B(x \rightarrow y)$ and $\Gamma(x \rightarrow y)$ denotes the branching ratios and the resonance width of the process $x \rightarrow y$.

The ratio of tau and muon lifetime is thus given by

$$\frac{\tau_\tau}{\tau_\mu} = \frac{B\left(\tau^- \rightarrow e^- + \overline{v}_e + v_\tau\right)}{B\left(\mu^- \rightarrow e^- + \overline{v}_e + v_\mu\right)}\left(\frac{m_\mu}{m_\tau}\right)^5.$$

Using the values of the 2008 *Review of Particle Physics* for the branching ratios of muons and tau yields a lifetime ratio of ~1.29×10^{-7}, comparable to the measured lifetime ratio of ~1.32×10^{-7}. The difference is due to K_1 and K_2 not actually being constants; they depend on the mass of leptons.

Table of Leptons

Properties of leptons									
Particle/anti-particle name	Symbol	Q (e)	S	L_e	L_μ	L_τ	Mass (MeV/c²)	Lifetime (s)	Common decay
Electron / Positron	e− / e+	−1 / +1	½	+1 / −1	0	0	0.510998910(13)	Stable	Stable
Muon / Antimuon	μ− / μ+	−1 / +1	½	0	+1 / −1	0	105.6583668(38)	2.197019(21)×10⁻⁶	e− + v e + v μ
Tau / Antitau	τ− / τ+	−1 / +1	½	0	0	+1 / −1	1776.84(17)	2.906(10)×10⁻¹³	See τ− decay modes
Electron neutrino / Electron antineutrino	ve / ve	0	½	+1 / −1	0	0	< 0.0000022	Unknown	
Muon neutrino / Muon antineutrino	vμ / vμ	0	½	0	+1 / −1	0	< 0.17	Unknown	
Tau neutrino / Tau antineutrino	vτ / vτ	0	½	0	0	+1 / −1	< 15.5	Unknown	

Gluon

Gluons are elementary particles that act as the exchange particles (or gauge bosons) for the strong force between quarks, analogous to the exchange of photons in the electromagnetic force between two charged particles. In layman terms, they "glue" quarks together, forming protons and neutrons.

In technical terms, gluons are vector gauge bosons that mediate strong interactions of quarks in quantum chromodynamics (QCD). Gluons themselves carry the color charge of the strong interaction. This is unlike the photon, which mediates the electromagnetic interaction but lacks an electric charge. Gluons therefore participate in the strong interaction in addition to mediating it, making QCD significantly harder to analyze than QED (quantum electrodynamics).

Properties

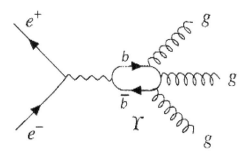

Diagram 2: $e^+e^- \rightarrow Y(9.46) \rightarrow 3g$

The gluon is a vector boson; like the photon, it has a spin of 1. While massive spin-1 particles have three polarization states, massless gauge bosons like the gluon have only two polarization states because gauge invariance requires the polarization to be transverse. In quantum field theory, unbroken gauge invariance requires that gauge bosons have zero mass (experiment limits the gluon's rest mass to less than a few meV/c^2). The gluon has negative intrinsic parity.

Numerology of Gluons

Unlike the single photon of QED or the three W and Z bosons of the weak interaction, there are eight independent types of gluon in QCD.

This may be difficult to understand intuitively. Quarks carry three types of color charge; antiquarks carry three types of anticolor. Gluons may be thought of as carrying both color and anticolor, but to correctly understand how they are combined, it is necessary to consider the mathematics of color charge in more detail.

Color Charge and Superposition

In quantum mechanics, the states of particles may be added according to the principle of superposition; that is, they may be in a "combined state" with a *probability*, if some particular quantity is measured, of giving several different outcomes. A relevant illustration in the case at hand would be a gluon with a color state described by:

$$(r\bar{b} + b\bar{r}) / \sqrt{2}.$$

This is read as "red–antiblue plus blue–antired". (The factor of the square root of two is required for normalization, a detail that is not crucial to understand in this discussion.) If one were somehow able to make a direct measurement of the color of a gluon in this state, there would be a 50% chance of it having red-antiblue color charge and a 50% chance of blue-antired color charge.

Color Singlet States

It is often said that the stable strongly interacting particles (such as the proton and the neutron, i.e. hadrons) observed in nature are "colorless", but more precisely they are in a "color singlet" state, which is mathematically analogous to a *spin* singlet state. Such states allow interaction with other color singlets, but not with other color states; because long-range gluon interactions do not exist, this illustrates that gluons in the singlet state do not exist either.

The color singlet state is:

$$(r\bar{r} + b\bar{b} + g\bar{g}) / \sqrt{3}.$$

In words, if one could measure the color of the state, there would be equal probabilities of it being red-antired, blue-antiblue, or green-antigreen.

Eight Gluon Colors

There are eight remaining independent color states, which correspond to the "eight types" or "eight colors" of gluons. Because states can be mixed together as discussed above, there are many ways of presenting these states, which are known as the "color octet". One commonly used list is:

$$(r\bar{b} + b\bar{r}) / \sqrt{2} \quad -i(r\bar{b} - b\bar{r}) / \sqrt{2}$$
$$(r\bar{g} + g\bar{r}) / \sqrt{2} \quad -i(r\bar{g} - g\bar{r}) / \sqrt{2}$$
$$(b\bar{g} + g\bar{b}) / \sqrt{2} \quad -i(b\bar{g} - g\bar{b}) / \sqrt{2}$$
$$(r\bar{r} - b\bar{b}) / \sqrt{2} \quad (r\bar{r} + b\bar{b} - 2g\bar{g}) / \sqrt{6}.$$

These are equivalent to the Gell-Mann matrices. The critical feature of these particular eight states is that they are linearly independent, and also independent of the singlet state; there is no way to add any combination of states to produce any other. (It is also impossible to add them to make rr, gg, or bb otherwise the forbidden singlet state could also be made.) There are many other possible choices, but all are mathematically equivalent, at least equally complex, and give the same physical results.

Group Theory Details

Technically, QCD is a gauge theory with SU(3) gauge symmetry. Quarks are introduced as spinors in N_f flavors, each in the fundamental representation (triplet, denoted 3) of the color gauge group, SU(3). The gluons are vectors in the adjoint representation (octets, denoted 8) of color SU(3). For a general gauge group, the number of force-carriers (like photons or gluons) is always equal to the dimension of the adjoint representation. For the simple case of SU(N), the dimension of this representation is $N^2 - 1$.

In terms of group theory, the assertion that there are no color singlet gluons is simply the statement that quantum chromodynamics has an SU(3) rather than a U(3) symmetry. There is no known *a priori* reason for one group to be preferred over the other, but as discussed above, the experimental evidence supports SU(3). The U(1) group for electromagnetic field combines with a slightly more complicated group known as SU(2) – S stands for "special" – which means the corresponding matrices have determinant 1 in addition to being unitary.

Confinement

Since gluons themselves carry color charge, they participate in strong interactions. These gluon-gluon interactions constrain color fields to string-like objects called "flux tubes", which exert constant force when stretched. Due to this force, quarks are confined within composite particles called hadrons. This effectively limits the range of the strong interaction to 1×10^{-15} meters, roughly the size of an atomic nucleus. Beyond a certain distance, the energy of the flux tube binding two quarks increases linearly. At a large enough distance, it becomes energetically more favorable to pull a quark-antiquark pair out of the vacuum rather than increase the length of the flux tube.

Gluons also share this property of being confined within hadrons. One consequence is that gluons are not directly involved in the nuclear forces between hadrons. The force mediators for these are other hadrons called mesons.

Although in the normal phase of QCD single gluons may not travel freely, it is predicted that there exist hadrons that are formed entirely of gluons — called glueballs. There are also conjectures about other exotic hadrons in which real gluons (as opposed to virtual ones found in ordinary hadrons) would be primary constituents. Beyond the normal phase of QCD (at extreme temperatures and pressures), quark–gluon plasma forms. In such a plasma there are no hadrons; quarks and gluons become free particles.

Experimental Observations

Quarks and gluons (colored) manifest themselves by fragmenting into more quarks and gluons, which in turn hadronize into normal (colorless) particles, correlated in jets. As shown in 1978 summer conferences the PLUTO detector at the electron-positron collider DORIS (DESY) produced the first evidence that the hadronic decays of the very narrow resonance Y(9.46) could be interpreted as three-jet event topologies produced by three gluons. Later published analyses by the same experiment confirmed this interpretation and also the spin 1 nature of the gluon.

In summer 1979 at higher energies at the electron-positron collider PETRA (DESY) again three-jet topologies were observed, now interpreted as qq gluon bremsstrahlung, now clearly visible, by TASSO, MARK-J and PLUTO experiments (later in 1980 also by JADE). The spin 1 of the gluon was confirmed in 1980 by TASSO and PLUTO experiments. In 1991 a subsequent experiment at the LEP storage ring at CERN again confirmed this result.

The gluons play an important role in the elementary strong interactions between quarks and gluons, described by QCD and studied particularly at the electron-proton collider HERA at DESY. The number and momentum distribution of the gluons in the proton (gluon density) have been measured by two experiments, H1 and ZEUS, in the years 1996 till today (2012). The gluon contribution to the proton spin has been studied by the HERMES experiment at HERA. The gluon density in the proton (when behaving hadronically) also has been measured.

Color confinement is verified by the failure of free quark searches (searches of fractional charges). Quarks are normally produced in pairs (quark + antiquark) to compensate the quantum color and flavor numbers; however at Fermilab single production of top quarks has been shown (technically this still involves a pair production, but quark and antiquark are of different flavor). No glueball has been demonstrated.

Deconfinement was claimed in 2000 at CERN SPS in heavy-ion collisions, and it implies a new state of matter: quark–gluon plasma, less interacting than in the nucleus, almost as in a liquid. It was found at the Relativistic Heavy Ion Collider (RHIC) at Brookhaven in the years 2004–2010 by four contemporaneous experiments. A quark–gluon plasma state has been confirmed at the CERN Large Hadron Collider (LHC) by the three experiments ALICE, ATLAS and CMS in 2010.

The Continuous Electron Beam Accelerator Facility at Jefferson Lab, also called the Thomas Jefferson National Accelerator Facility, in Newport News, Virginia is one of 10 Department of Energy facilities doing research on gluons. The Virginia lab is competing with another facility in Long Island, New York, Brookhaven National Laboratory, for funds to build a new electron-ion collider.

W and Z Bosons

The W and Z bosons are together known as the weak or more generally as the intermediate vector bosons. These elementary particles mediate the weak interaction; the respective symbols are W+, W–, and Z. The W boson have either a positive and negative electric charge of 1 elementary charge and are each other's antiparticles. The Z boson is electrically neutral and is its own antiparticle. The three particles have a spin of 1. The W bosons have a magnetic moment, but the Z has none. All three of these particles are very short-lived, with a half-life of about 3×10^{-25} s. Their experimental discovery was a triumph for what is now known as the Standard Model of particle physics.

The W bosons are named after the *weak* force. The physicist Steven Weinberg named the additional particle the "Z particle", and later gave the explanation that it was the last additional particle needed by the model. The W bosons had already been named, and the Z bosons have *zero* electric charge.

The two W bosons are verified mediators of neutrino absorption and emission. During these processes, the W boson charge induces electron or positron emission or absorption, thus causing nuclear transmutation. The Z boson is not involved in the absorption or emission of electrons and positrons.

The Z boson mediates the transfer of momentum, spin and energy when neutrinos scatter *elastically* from matter (a process which conserves charge). Such behavior is almost as common as inelastic neutrino interactions and may be observed in bubble chambers upon irradiation with neutrino beams. Whenever an electron is observed as a new free particle suddenly moving with kinetic energy, it is inferred to be a result of a neutrino interacting directly with the electron if this behavior happens more often when the neutrino beam is present. In this process, the neutrino simply strikes the electron and then scatters away from it, transferring some of the neutrino's momentum to the electron. Because neutrinos are neither affected by the strong force nor the electromagnetic force, and because the gravitational force between subatomic particles is negligible, such an interaction can only happen via the weak force. Since such an electron is not created from a nucleon, and is unchanged except for the new force impulse imparted by the neutrino, this weak force interaction between the neutrino and the electron must be mediated by an electromagnetically neutral, weak-force boson particle. Thus, this interaction requires a Z boson.

Basic Properties

These bosons are among the heavyweights of the elementary particles. With masses of 80.4 GeV/c^2 and 91.2 GeV/c^2, respectively, the W and Z bosons are almost 100 times as large as the proton – heavier, even, than entire iron atoms. The masses of these bosons are significant because they act as the force carriers of a quite short-range fundamental force: their high masses thus limit the range of the weak nuclear force. By way of contrast, the electromagnetic force has an infinite range, because its force carrier, the photon, has zero mass, and the same is supposed of the hypothetical graviton.

All three bosons have particle spin $s = 1$. The emission of a W+ or W– boson either raises or lowers the electric charge of the emitting particle by one unit, and also alters the spin by one unit. At the same time, the emission or absorption of a W boson can change the type of the particle – for example changing a strange quark into an up quark. The neutral Z boson cannot change the electric charge of any particle, nor can it change any other of the so-called "charges" (such as strangeness, baryon number, charm, etc.). The emission or absorption of a Z boson can only change the spin, momentum, and energy of the other particle.

Weak Nuclear Force

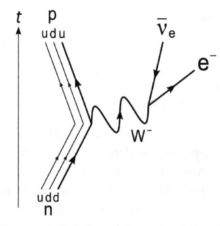

The Feynman diagram for beta decay of a neutron into a proton, electron, and electron antineutrino via an intermediate heavy W boson

The W and Z bosons are carrier particles that mediate the weak nuclear force, much as the photon is the carrier particle for the electromagnetic force.

W bosons

The W bosons are best known for their role in nuclear decay. Consider, for example, the beta decay of cobalt-60.

$$6027Co \rightarrow 6028Ni^+ + e- + ve$$

This reaction does not involve the whole cobalt-60 nucleus, but affects only one of its 33 neutrons. The neutron is converted into a proton while also emitting an electron (called a beta particle in this context) and an electron antineutrino:

$$no \rightarrow p+ + e- + ve$$

Again, the neutron is not an elementary particle but a composite of an up quark and two down quarks (udd). It is in fact one of the down quarks that interacts in beta decay, turning into an up quark to form a proton (uud). At the most fundamental level, then, the weak force changes the flavour of a single quark:

$$d \rightarrow u + W-$$

which is immediately followed by decay of the W− itself:

$$W- \rightarrow e- + ve$$

Z Boson

The Z boson is its own antiparticle. Thus, all of its flavour quantum numbers and charges are zero. The exchange of a Z boson between particles, called a neutral current interaction, therefore leaves the interacting particles unaffected, except for a transfer of momentum. Z boson interactions involving neutrinos have distinctive signatures: They provide the only known mechanism for elastic scattering of neutrinos in matter; neutrinos are almost as likely to scatter elastically (via Z boson exchange) as inelastically (via W boson exchange). The first prediction of Z bosons was made by Brazilian physicist José Leite Lopes in 1958, by devising an equation which showed the analogy of the weak nuclear interactions with electromagnetism. Steve Weinberg, Sheldon Glashow and Abdus Salam used later these results to develop the electroweak unification, in 1973. Weak neutral currents via Z boson exchange were confirmed shortly thereafter in 1974, in a neutrino experiment in the Gargamelle bubble chamber at CERN.

Predicting the W and Z

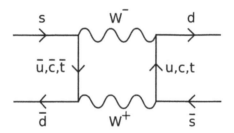

A Feynman diagram showing the exchange of a pair of W bosons.
This is one of the leading terms contributing to neutral Kaon oscillation.

Following the spectacular success of quantum electrodynamics in the 1950s, attempts were undertaken to formulate a similar theory of the weak nuclear force. This culminated around 1968 in a unified theory of electromagnetism and weak interactions by Sheldon Glashow, Steven Weinberg, and Abdus Salam, for which they shared the 1979 Nobel Prize in Physics. Their electroweak theory postulated not only the W bosons necessary to explain beta decay, but also a new Z boson that had never been observed.

The fact that the W and Z bosons have mass while photons are massless was a major obstacle in developing electroweak theory. These particles are accurately described by an SU(2) gauge theory, but the bosons in a gauge theory must be massless. As a case in point, the photon is massless because electromagnetism is described by a U(1) gauge theory. Some mechanism is required to break the SU(2) symmetry, giving mass to the W and Z in the process. One explanation, the Higgs mechanism, was forwarded by the 1964 PRL symmetry breaking papers. It predicts the existence of yet another new particle; the Higgs boson. Of the four components of a Goldstone boson created by the Higgs field, three are "eaten" by the W^+, Z^0, and W^- bosons to form their longitudinal components and the remainder appears as the spin 0 Higgs boson.

The combination of the SU(2) gauge theory of the weak interaction, the electromagnetic interaction, and the Higgs mechanism is known as the Glashow-Weinberg-Salam model. These days it is widely accepted as one of the pillars of the Standard Model of particle physics. As of 13 December 2011, intensive search for the Higgs boson carried out at CERN has indicated that if the particle is to be found, it seems likely to be found around 125 GeV. On 4 July 2012, the CMS and the ATLAS experimental collaborations at CERN announced the discovery of a new particle with a mass of 125.3 ± 0.6 GeV that appears consistent with a Higgs boson.

Discovery

The Gargamelle bubble chamber, now exhibited at CERN

Unlike beta decay, the observation of neutral current interactions that involve particles *other than neutrinos* requires huge investments in particle accelerators and detectors, such as are available in only a few high-energy physics laboratories in the world (and then only after 1983). This is because Z-bosons behave in somewhat the same manner as photons, but do not become important until the energy of the interaction is comparable with the relatively huge mass of the Z boson.

The discovery of the W and Z bosons was considered a major success for CERN. First, in 1973, came the observation of neutral current interactions as predicted by electroweak theory. The huge Gargamelle bubble chamber photographed the tracks of a few electrons suddenly starting to move, seemingly of their own accord. This is interpreted as a neutrino interacting with the electron by the exchange of an unseen Z boson. The neutrino is otherwise undetectable, so the only observable effect is the momentum imparted to the electron by the interaction.

The discovery of the W and Z bosons themselves had to wait for the construction of a particle accelerator powerful enough to produce them. The first such machine that became available was the Super Proton Synchrotron, where unambiguous signals of W bosons were seen in January 1983 during a series of experiments made possible by Carlo Rubbia and Simon van der Meer. The actual experiments were called UA1 (led by

Rubbia) and UA2 (led by Pierre Darriulat), and were the collaborative effort of many people. Van der Meer was the driving force on the accelerator end (stochastic cooling). UA1 and UA2 found the Z boson a few months later, in May 1983. Rubbia and van der Meer were promptly awarded the 1984 Nobel Prize in Physics, a most unusual step for the conservative Nobel Foundation.

The W+, W−, and Z0 bosons, together with the photon (γ), comprise the four gauge bosons of the electroweak interaction.

Decay

The W and Z bosons decay to fermion–antifermion pairs but neither the W nor the Z bosons can decay into the higher-mass top quark. Neglecting phase space effects and higher order corrections, simple estimates of their branching fractions can be calculated from the coupling constants.

W Bosons

W bosons can decay to a lepton and neutrino or to an up-type quark and a down-type quark. The decay width of the W boson to a quark–antiquark pair is proportional to the corresponding squared CKM matrix element and the number of quark colours, $N_C = 3$. The decay widths for the W bosons are then proportional to:

Leptons		Up quarks		Charm quarks	
e+ve	1	ud	$3\lvert V_{ud}\rvert^2$	cd	$3\lvert V_{cd}\rvert^2$
μ+vμ	1	us	$3\lvert V_{us}\rvert^2$	cs	$3\lvert V_{cs}\rvert^2$
τ+vτ	1	ub	$3\lvert V_{ub}\rvert^2$	cb	$3\lvert V_{cb}\rvert^2$

Here, e+, μ+, τ+ denote the three flavours of leptons (more exactly, the positive charged antileptons). ve, vμ, vτ denote the three flavours of neutrinos. The other particles, starting with u and d, all denote quarks and antiquarks (factor N_C is applied). The various V_{ij} denote the corresponding CKM matrix coefficients.

Unitarity of the CKM matrix implies that $\lvert V_{ud}\rvert^2 + \lvert V_{us}\rvert^2 + \lvert V_{ub}\rvert^2 = \lvert V_{cd}\rvert^2 + \lvert V_{cs}\rvert^2 + \lvert V_{cb}\rvert^2 = 1$. Therefore, the leptonic branching ratios of the W boson are approximately $B(e+ve) = B(\mu+v\mu) = B(\tau+v\tau) = \frac{1}{9}$. The hadronic branching ratio is dominated by the CKM-favored ud and cs final states. The sum of the hadronic branching ratios has been measured experimentally to be 67.60±0.27%, with $B(l^+v_l) = 10.80\pm0.09\%$.

Z Bosons

Z bosons decay into a fermion and its antiparticle. As the Z-boson is a mixture of the pre-symmetry-breaking W^0 and B^0 bosons, each vertex factor includes a factor $T_3 - Qsin^2\theta_W$, where T_3 is the third component of the weak isospin of the fermion, Q is the electric charge of the fermion (in units of the elementary charge), and θ_W is the weak mixing angle. Because the weak isospin is different for fermions of different chirality, either left-handed or right-handed, the coupling is different as well.

The relative strengths of each coupling can be estimated by considering that the decay rates include the square of these factors, and all possible diagrams (e.g. sum over quark families, and left and right contributions). This is just an estimate, as we are considering only tree-level diagrams in the Fermi theory.

Particles		Effective charge		Relative factor	Branching ratio	
Name	Symbols	L	R		Predicted for $x = 0.23$	Experimental measurements
Neutrinos (all)	νe, νμ, ντ	$\frac{1}{2}$	0	$3(\frac{1}{2})^2$	20.5%	20.00±0.06%
Charged leptons (all)	e-, μ-, τ-			$3((-\frac{1}{2}+x)^2 + x^2)$	10.2%	10.097±0.003%
Electron	e-	$-\frac{1}{2}+x$	x	$(-\frac{1}{2}+x)^2 + x^2$	3.4%	3.363±0.004%
Muon	μ-	$-\frac{1}{2}+x$	x	$(-\frac{1}{2}+x)^2 + x^2$	3.4%	3.366±0.007%
Tau	τ-	$-\frac{1}{2}+x$	x	$(-\frac{1}{2}+x)^2 + x^2$	3.4%	3.367±0.008%
Hadrons (all)					69.2%	69.91±0.06%
Down-type quarks	d, s, b	$-\frac{1}{2}+\frac{1}{3}x$	$\frac{1}{3}x$	$3(-\frac{1}{2}+\frac{1}{3}x)^2 + 3(\frac{1}{3}x)^2$	15.2%	15.6±0.4%
Up-type quarks	u, c	$\frac{1}{2}-\frac{2}{3}x$	$-\frac{2}{3}x$	$3(\frac{1}{2}-\frac{2}{3}x)^2 + 3(-\frac{2}{3}x)^2$	11.8%	11.6±0.6%

Here, L and R denote either the left- or right-handed chirality of the fermions respectively. (The right-handed neutrinos do not exist in the standard model. However, in some extensions beyond the standard model they do.) The notation $x = sin^2\theta_W$ is used.

Higgs Boson

The Higgs boson is an elementary particle in the Standard Model of particle physics. It is the quantum excitation of the Higgs field, a fundamental field of crucial importance to particle physics theory first suspected to exist in the 1960s. Unlike other known fields such as the electromagnetic field, it takes a non-zero constant value almost everywhere. The question of the Higgs field's existence has been the last unverified part of the Standard Model of particle physics and, according to some, "the central problem in particle physics".

The presence of this field, now believed to be confirmed, explains why some fundamental particles have mass when, based on the symmetries controlling their interactions, they should be massless. The existence of the Higgs field would also resolve several other long-standing puzzles, such as the reason for the weak force's extremely short range.

Although it is hypothesized that the Higgs field permeates the entire Universe, evidence for its existence has been very difficult to obtain. In principle, the Higgs field can be detected through its excitations, manifest as Higgs particles, but these are extremely difficult to produce and detect. The importance of this fundamental question led to a 40 year search, and the construction of one of the world's most expensive and complex experimental facilities to date, CERN's Large Hadron Collider, in an attempt to create Higgs bosons and other particles for observation and study. On 4 July 2012, the discovery of a new particle with a mass between 125 and 127 GeV/c^2 was announced; physicists suspected that it was the Higgs boson. Since then, the particle has been shown to behave, interact, and decay in many of the ways predicted by the Standard Model. It was also tentatively confirmed to have even parity and zero spin, two fundamental attributes of a Higgs boson. This appears to be the first elementary scalar particle discovered in nature. More studies are needed to verify that the discovered particle has properties matching those predicted for the Higgs boson by the Standard Model, or whether, as predicted by some theories, multiple Higgs bosons exist.

The Higgs boson is named after Peter Higgs, one of six physicists who, in 1964, proposed the mechanism that suggested the existence of such a particle. On December 10, 2013, two of them, Peter Higgs and François Englert, were awarded the Nobel Prize in Physics for their work and prediction (Englert's co-researcher Robert Brout had died in 2011 and the Nobel Prize is not ordinarily given posthumously). Although Higgs's name has come to be associated with this theory, several researchers between about 1960 and 1972 independently developed different parts of it. In mainstream media the Higgs boson has often been called the "God particle", from a 1993 book on the topic; the nickname is strongly disliked by many physicists, including Higgs, who regard it as sensationalistic.

In the Standard Model, the Higgs particle is a boson with no spin, electric charge, or colour charge. It is also very unstable, decaying into other particles almost immediately. It is a quantum excitation of one of the four components of the Higgs field. The latter constitutes a scalar field, with two neutral and two electrically charged components that form a complex doublet of the weak isospin SU(2) symmetry. The Higgs field is tachyonic (this does not refer to faster-than-light speeds, it means that symmetry-breaking through condensation of a particle must occur under certain conditions), and has a "Mexican hat" shaped potential with nonzero strength everywhere (including otherwise empty space), which in its vacuum state breaks the weak isospin symmetry of the electroweak interaction. When this happens, three components of the Higgs field are "absorbed" by the SU(2) and U(1) gauge bosons (the "Higgs mechanism") to become the longitudinal components of the now-massive W and Z bosons of the weak force. The remaining electrically neutral component separately couples to other particles known as fermions (via Yukawa couplings), causing these to acquire mass as well. Some versions of the theory predict more than one kind of Higgs fields and bosons. Alternative "Higgs-less" models may have been considered if the Higgs boson was not discovered.

A Non-technical Summary

"Higgs" Terminology

A simple explanation – what are the Higgs mechanism, field and boson?	
Gauge symmetries	In the Standard Model of particle physics, the fundamental forces of nature arise from laws of nature called gauge symmetries, and these forces are transmitted by particles known as gauge bosons. According to the gauge theory, the weak force's gauge symmetry should cause its gauge bosons to have zero mass, but experiments show that the weak force's gauge bosons, the W and Z, have mass. But it has proved difficult to find any way to explain their unexpected mass.
Higgs mechanism	By the early 1960s, physicists had realized that a given symmetry law might not always be followed (or 'obeyed') under certain conditions. The Higgs mechanism is a mathematical model devised by three groups of researchers in 1964 that explains why and how gauge bosons could still be massive despite their governing symmetry. It showed that the conditions for the symmetry would be 'broken' if an unusual type of field existed throughout the Universe, and then why some fundamental particles would be able to have mass.
Higgs field	According to the Standard Model, a field of the necessary kind (the "Higgs field") exists throughout space, and breaks certain symmetry laws of the electroweak interaction. The existence of this field triggers the Higgs mechanism, causing the gauge bosons responsible for the weak force to be massive, and explaining their very short range. Some years after the original theory was articulated scientists realised that the same field would also explain, in a different way, why other fundamental constituents of matter (including electrons and quarks) have mass. For many years scientists had no way to tell whether or not the Higgs field existed. If it did, then it would be unlike other fundamental fields. But it was also possible that these key ideas, or even the entire Standard Model itself, were somehow incorrect. Only discovering what was breaking this symmetry would solve the problem.
Higgs boson	The existence of the Higgs field could be supported by searching for a matching particle associated with it — the "Higgs boson". Detecting Higgs bosons would prove that the Higgs field exists, and further support the validity of the Standard Model. But for decades scientists had no way to discover whether Higgs bosons actually existed in nature either, because they would be very difficult to produce, and would break apart in about a ten-sextillionth (10^{-22}) of a second. Although the theory gave "remarkably" accurate predictions, particle colliders, detectors, and computers capable of looking for Higgs bosons took over 30 years (c. 1980 – 2010) to develop. As of March 2016, scientists are confident that they have discovered the existence of the Higgs boson, and therefore the concept of some type of Higgs field throughout space is highly supported. Further testing over the coming years should eventually tell us more about these, and is likely to have significant impact in the future.

Overview

Physicists explain the properties and forces between elementary particles in terms of the Standard Model—a widely accepted and "remarkably" accurate framework based on gauge invariance and symmetries, believed to explain almost everything in the known universe, other than gravity. But by around 1960 all attempts to create a gauge invariant theory for two of the four fundamental forces had consistently failed at one crucial point: although gauge invariance seemed extremely important, it seemed to make any theory of electromagnetism and the weak force go haywire, by demanding that either many particles with mass were massless or that non-existent forces and massless particles had to exist. Scientists had no idea how to get past this point.

In 1962 physicist Philip Anderson wrote a paper that built upon work by Yoichiro Nambu concerning "broken symmetries" in superconductivity and particle physics. He suggested that "broken symmetries" might also be the missing piece needed to solve the problems of gauge invariance. In 1964 a theory was created almost simultaneously by 3 different groups of researchers, that showed Anderson's suggestion was possible - the gauge theory and "mass problems" could indeed be resolved if an unusual kind of field, now generally called the "Higgs field", existed throughout the universe; if the Higgs field did exist, it would apparently cause existing particles to acquire mass instead of new massless particles being formed. Although these ideas did not gain much initial support or attention, by 1972 they had been developed into a comprehensive theory and proved capable of giving "sensible" results that accurately described particles known at the time, and which accurately predicted several other particles discovered during the following years. During the 1970s these theories rapidly became the "standard model". There was not yet any direct evidence that the Higgs field actually existed, but even without proof of the field, the accuracy of its predictions led scientists to believe the theory might be true. By the 1980s the question whether or not the Higgs field existed had come to be regarded as one of the most important unanswered questions in particle physics.

If the Higgs field could be shown to exist, it would be a monumental discovery for science and human knowledge, and would open doorways to new knowledge in many disciplines. If not, then other more complicated theories would need to be considered. The simplest means to test the existence of the Higgs field would be a search for a new elementary particle that the field would have to give off, a particle known as the "Higgs boson" or the "Higgs particle". This particle would be extremely difficult to find. After significant technological advancements, by the 1990s two large experimental installations were being designed and constructed that allowed to search for the Higgs boson.

While several symmetries in nature are spontaneously broken through a form of the Higgs mechanism, in the context of the Standard Model the term "Higgs mechanism" almost always means symmetry breaking of the electroweak field. It is considered confirmed, but revealing the exact cause has been difficult. Various analogies have also been invented to describe the Higgs field and boson, including analogies with well-known symmetry breaking effects such as the rainbow and prism, electric fields, ripples, and resistance of macro objects moving through media, like people moving through crowds or some objects moving through syrup or molasses. However, analogies based on simple resistance to motion are inaccurate as the Higgs field does not work by resisting motion.

Significance

Scientific Impact

Evidence of the Higgs field and its properties has been extremely significant scientifically, for many reasons. The Higgs boson's importance is largely that it is able to be examined using existing knowledge and experimental technology, as a way to confirm and study the entire Higgs field theory. Conversely, proof that the Higgs field and boson do *not* exist would also have been significant. In discussion form, the relevance includes:

Validating the Standard Model, or choosing between extensions and alternatives	Does the Higgs field exist, which fundamentally validates the Standard Model through the mechanism of Mass generation? If it does, then which more advanced extensions are suggested or excluded based upon measurements of its properties? What else can we learn about this fundamental field, now that we have the experimental means to study its behavior and interactions? Alternatively, if the Higgs field doesn't exist, which alternatives and modifications to the Standard Model are likely to be preferred? Will the data suggest an extension, or a completely different approach (such as supersymmetry or string theory)?
	Related to this, a belief generally exists among physicists that there is likely to be "new" physics beyond the Standard Model—the Standard Model will at some point be extended or superseded. The Higgs field and related issues present a promising "doorway" to understand better the places where the Standard Model might become inadequate or fail, and could provide considerable evidence guiding researchers into future enhancements or successors.
Finding how symmetry breaking happens within the electroweak interaction	Below an extremely high temperature, electroweak symmetry breaking causes the electroweak interaction to manifest in part as the short-ranged weak force, which is carried by massive gauge bosons. Without this, the universe we see around us could not exist, because atoms and other structures could not form, and reactions in stars such as our Sun would not occur. But it is not clear how this actually happens in nature. Is the Standard Model correct in its approach, and can it be made more exact with actual experimental measurements? If not the Higgs field, then what is breaking symmetry in its place?
Finding how certain particles acquire mass	Electroweak symmetry breaking (due to a Higgs field or otherwise) is believed proven responsible for the masses of fundamental particles such as elementary fermions (including electrons and quarks) and the massive W and Z gauge bosons. Finding how this happens is pivotal to particle physics.
	It is worth noting that the Higgs field does not 'create' mass out of nothing (which would violate the law of conservation of energy). Nor is the Higgs field responsible for the mass of all particles. For example, about 99% of the mass of baryons (composite particles such as the proton and neutron) is due instead to the kinetic energy of quarks and to the energies of (massless) gluons of the strong interaction inside the baryons. In Higgs-based theories, the property of 'mass' is a manifestation of potential energy transferred to particles when they interact ("couple") with the Higgs field, which had contained that mass in the form of energy.

Evidence whether or not scalar fields exist in nature, and "new" physics	The significance of proof of a scalar field such as the Higgs field would be hard to over-estimate: "[The] verification of real scalar fields would be nearly as important as its role in generating mass". Rolf-Dieter Heuer, director general of CERN, stated in a 2011 talk on the Higgs field: All the matter particles are spin-1/2 fermions. All the force carriers are spin-1 bosons. Higgs particles are spin-0 bosons (scalars). The Higgs is neither matter nor force. The Higgs is just different. This would be the first fundamental scalar ever discovered. The Higgs field is thought to fill the entire universe. Could it give some handle of dark energy (scalar field)? Many modern theories predict other scalar particles like the Higgs. Why, after all, should the Higgs be the only one of its kind? [The] LHC can search for and study new scalars with precision.
Insight into cosmic inflation	There has been considerable scientific research on possible links between the Higgs field and the inflaton – a hypothetical field suggested as the explanation for the expansion of space during the first fraction of a second of the universe (known as the "inflationary epoch"). Some theories suggest that a fundamental scalar field might be responsible for this phenomenon; the Higgs field is such a field and therefore has led to papers analysing whether it could also be the *inflaton* responsible for this exponential expansion of the universe during the Big Bang. Such theories are highly tentative and face significant problems related to unitarity, but may be viable if combined with additional features such as large non-minimal coupling, a Brans–Dicke scalar, or other "new" physics, and have received treatments suggesting that Higgs inflation models are still of interest theoretically.
Insight into the nature of the universe, and its possible fates	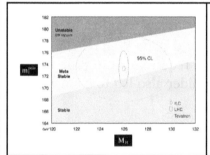 Diagram showing the Higgs boson and top quark masses, which could indicate whether our universe is stable, or a long-lived 'bubble'. As of 2012, the 2σ ellipse based on Tevatron and LHC data still allows for both possibilities. For decades, scientific models of our universe have included the possibility that it exists as a long-lived, but not completely stable, sector of space, which could potentially at some time be destroyed upon 'toppling' into a more stable vacuum state. If the masses of the Higgs boson and top quark are known more exactly, and the Standard Model provides an accurate description of particle physics up to extreme energies of the Planck scale, then it is possible to calculate whether the universe's present vacuum state is stable or merely long-lived. (This was sometimes misreported as the Higgs boson "ending" the universe). A 125 – 127 GeV Higgs mass seems to be extremely close to the boundary for stability (estimated in 2012 as 123.8 – 135.0 GeV) but a definitive answer requires much more precise measurements of the top quark's pole mass. New physics can change this picture. If measurements of the Higgs boson suggest that our universe lies within a false vacuum of this kind, then it would imply – more than likely in many billions of years – that the universe's forces, particles, and structures could cease to exist as we know them (and be replaced by different ones), if a true vacuum happened to nucleate. It also suggests that the Higgs self-coupling λ and its β_λ function could be very close to zero at the Planck scale, with "intriguing" implications, including theories of gravity and Higgs-based inflation. A future electron–positron collider would be able to provide the precise measurements of the top quark needed for such calculations.

Insight into the 'energy of the vacuum'	More speculatively, the Higgs field has also been proposed as the energy of the vacuum, which at the extreme energies of the first moments of the Big Bang caused the universe to be a kind of featureless symmetry of undifferentiated extremely high energy. In this kind of speculation, the single unified field of a Grand Unified Theory is identified as (or modeled upon) the Higgs field, and it is through successive symmetry breakings of the Higgs field or some similar field at phase transitions that the present universe's known forces and fields arise.
Link to the 'cosmological constant' problem	The relationship (if any) between the Higgs field and the presently observed vacuum energy density of the universe has also come under scientific study. As observed, the present vacuum energy density is extremely close to zero, but the energy density expected from the Higgs field, supersymmetry, and other current theories are typically many orders of magnitude larger. It is unclear how these should be reconciled. This cosmological constant problem remains a further major unanswered problem in physics.

Practical and Technological Impact of Discovery

As yet, there are no known immediate technological benefits of finding the Higgs particle. However, a common pattern for fundamental discoveries is for practical applications to follow later, once the discovery has been explored further, at which point they become the basis for new technologies of importance to society.

The challenges in particle physics have furthered major technological progress of widespread importance. For example, the World Wide Web began as a project to improve CERN's communication system. CERN's requirement to process massive amounts of data produced by the Large Hadron Collider also led to contributions to the fields of distributed and cloud computing.

History

The six authors of the 1964 PRL papers, who received the 2010 J. J. Sakurai Prize for their work. From left to right: Kibble, Guralnik, Hagen, Englert, Brout. *Right:* Higgs.

Particle physicists study matter made from fundamental particles whose interactions are mediated by exchange particles - gauge bosons - acting as force carriers. At the beginning of the 1960s a number of these particles had been discovered or proposed, along with theories suggesting how they relate to each other, some of which had already been reformulated as field theories in which the objects of study are not particles and forces, but quantum fields and their symmetries. However, attempts to unify known fundamental forces such as the electromagnetic force and the weak nuclear force were known to be incomplete. One known omission was that gauge invariant approaches, including non-abelian models such as Yang–Mills theory (1954), which held great promise for unified theories, also seemed to predict known massive particles as massless. Goldstone's theorem, relating to continuous symmetries within some theories, also appeared to rule out many obvious solutions, since it appeared to show that zero-mass particles would have to also exist that were "simply not seen". According to Guralnik, physicists had "no understanding" how these problems could be overcome.

Nobel Prize Laureate Peter Higgs in Stockholm, December 2013

Particle physicist and mathematician Peter Woit summarised the state of research at the time:

Yang and Mills work on non-abelian gauge theory had one huge problem: in perturbation theory it has massless particles which don't correspond to anything we see. One way of getting rid of this problem is now fairly well-understood, the phenomenon of confinement realized in QCD, where the strong interactions get rid of the massless "gluon" states at long distances. By the very early sixties, people had begun to understand another source of massless particles: spontaneous symmetry breaking of a continuous symmetry. What Philip Anderson realized and worked out in the summer of 1962 was that, when you have *both* gauge symmetry *and* spontaneous symmetry breaking, the Nambu–Goldstone massless mode can combine with the massless gauge field modes to produce a physical massive vector field. This is what happens in superconductivity, a subject about which Anderson was (and is) one of the leading experts.

The Higgs mechanism is a process by which vector bosons can get rest mass *without* explicitly breaking gauge invariance, as a byproduct of spontaneous symmetry breaking. The mathematical theory behind spontaneous symmetry breaking was initially conceived and published within particle physics by Yoichiro Nambu in 1960, the concept that such a mechanism could offer a possible solution for the "mass problem" was originally suggested in 1962 by Philip Anderson (who had previously written papers on broken symmetry and its outcomes in superconductivity and concluded in his 1963 paper on Yang-Mills theory that "considering the superconducting analog... these two types of bosons seem capable of canceling each other out... leaving finite mass bosons"), and Abraham Klein and Benjamin Lee showed in March 1964 that Goldstone's theorem could be avoided this way in at least some non-relativistic cases and speculated it might be possible in truly relativistic cases.

These approaches were quickly developed into a full relativistic model, independently and almost simultaneously, by three groups of physicists: by François Englert and Robert Brout in August 1964; by Peter Higgs in October 1964; and by Gerald Guralnik, Carl Hagen, and Tom Kibble (GHK) in November 1964. Higgs also wrote a short but important response published in September 1964 to an objection by Gilbert, which showed that if calculating within the radiation gauge, Goldstone's theorem and Gilbert's objection would become inapplicable. (Higgs later described Gilbert's objection as prompting his own paper.) Properties of the model were further considered by Guralnik in 1965, by Higgs in 1966, by Kibble in 1967, and further by GHK in 1967. The original three 1964 papers showed that when a gauge theory is combined with an additional field that spontaneously breaks the symmetry, the gauge bosons can consistently acquire a finite mass. In 1967, Steven Weinberg and Abdus Salam independently showed how a Higgs mechanism could be used to break the electroweak symmetry of Sheldon Glashow's unified model for the weak and electromagnetic interactions (itself an extension of work by Schwinger), forming what became the Standard Model of particle physics. Weinberg was the first to observe that this would also provide mass terms for the fermions.

However, the seminal papers on spontaneous breaking of gauge symmetries were at first largely ignored, because it was widely believed that the (non-Abelian gauge) theories in question were a dead-end, and in particular that they could not be renormalised. In 1971–72, Martinus Veltman and Gerard 't Hooft proved renormalisation of Yang–Mills was possible in two papers covering massless, and then massive, fields. Their contribution, and others' work on the renormalization group - including "substantial" theoretical work by Russian physicists Ludvig Faddeev, Andrei Slavnov, Efim Fradkin and Igor Tyutin - was eventually "enormously profound and influential", but even with all key elements of the eventual theory published there was still almost no wider interest. For example, Coleman found in a study that "essentially no-one paid any attention" to Weinberg's paper prior to 1971 and discussed by David Politzer in his 2004 Nobel speech. – now the most cited in particle physics – and even in 1970 according to Politzer, Glashow's teaching of the weak interaction contained no mention of Weinberg's, Salam's, or Glashow's own work. In practice, Politzer states, almost everyone learned of the theory due to physicist Benjamin Lee, who combined the work of Veltman and 't Hooft with insights by others, and popularised the completed theory. In this way, from 1971, interest and acceptance "exploded" and the ideas were quickly absorbed in the mainstream.

The resulting electroweak theory and Standard Model have accurately predicted (among other things) weak neutral currents, three bosons, the top and charm quarks, and with great precision, the mass and other properties of some of these. Many of those involved eventually won Nobel Prizes or other renowned awards. A 1974 paper and comprehensive review in *Reviews of Modern Physics* commented that "while no one doubted the [mathematical] correctness of these arguments, no one quite believed that nature was diabolically clever enough to take advantage of them", adding that the theory had so far produced accurate answers that accorded with experiment, but it was unknown whether the theory was fundamentally correct. By 1986 and again in the 1990s it became possible to write that understanding and proving the Higgs sector of the Standard Model was "the central problem today in particle physics".

Summary and Impact of the PRL Papers

The three papers written in 1964 were each recognised as milestone papers during *Physical Review Letters*'s 50th anniversary celebration. Their six authors were also awarded the 2010 J. J. Sakurai Prize for Theoretical Particle Physics for this work. (A controversy also arose the same year, because in the event of a Nobel Prize only up to three scientists could be recognised, with six being credited for the papers.) Two of the three PRL papers (by Higgs and by GHK) contained equations for the hypothetical field that eventually would become known as the Higgs field and its hypothetical quantum, the Higgs boson. Higgs' subsequent 1966 paper showed the decay mechanism of the boson; only a massive boson can decay and the decays can prove the mechanism.

In the paper by Higgs the boson is massive, and in a closing sentence Higgs writes that "an essential feature" of the theory "is the prediction of incomplete multiplets of scalar and vector bosons". (Frank Close comments that 1960s gauge theorists were focused on the problem of massless *vector* bosons, and the implied existence of a massive *scalar* boson was not seen as important; only Higgs directly addressed it. In the paper by GHK the boson is massless and decoupled from the massive states. In reviews dated 2009 and 2011, Guralnik states that in the GHK model the boson is massless only in a lowest-order approximation, but it is not subject to any constraint and acquires mass at higher orders, and adds that the GHK paper was the only one to show that there are no massless Goldstone bosons in the model and to give a complete analysis of the general Higgs mechanism. All three reached similar conclusions, despite their very different approaches: Higgs' paper essentially used classical techniques, Englert and Brout's involved calculating vacuum polarization in perturbation theory around an assumed symmetry-breaking vacuum state, and GHK used operator formalism and conservation laws to explore in depth the ways in which Goldstone's theorem may be worked around.

Theoretical Properties

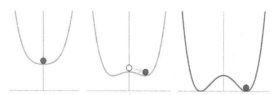

"Symmetry breaking illustrated": – At high energy levels *(left)* the ball settles in the center, and the result is symmetrical. At lower energy levels *(right)*, the overall "rules" remain symmetrical, but the "Mexican hat" potential comes into effect: "local" symmetry inevitably becomes broken since eventually the ball must at random roll one way or another.

Gauge invariance is an important property of modern particle theories such as the Standard Model, partly due to its success in other areas of fundamental physics such as electromagnetism and the strong interaction (quantum chromodynamics). However, there were great difficulties in developing gauge theories for the weak nuclear force or a possible unified electroweak interaction. Fermions with a mass term would violate gauge symmetry and therefore cannot be gauge invariant. (This can be seen by examining the Dirac Lagrangian for a fermion in terms of left and right handed components; we find none of the spin-half particles could ever flip helicity as required for mass, so they must be massless.[Note 13]) W and Z bosons are observed to have mass, but a boson mass term contains terms, which clearly depend on the choice of gauge and therefore these masses too cannot be gauge invariant. Therefore, it seems that *none* of the standard model fermions *or* bosons could "begin" with mass as an inbuilt property except by abandoning gauge invariance. If gauge invariance were to be retained, then these particles had to be acquiring their mass by some other mechanism or interaction. Additionally, whatever was giving these particles their mass, had to not "break" gauge invariance as the basis for other parts of the theories where it worked well, *and* had to not require or predict unexpected massless particles and long-range forces (seemingly an inevitable consequence of Goldstone's theorem) which did not actually seem to exist in nature.

A solution to all of these overlapping problems came from the discovery of a previously unnoticed borderline case hidden in the mathematics of Goldstone's theorem,[Note 11] that under certain conditions it *might* theoretically be possible for a symmetry to be broken *without* disrupting gauge invariance and *without* any new massless particles or forces, and having "sensible" (renormalisable) results mathematically: this became known as the Higgs mechanism.

The Standard Model hypothesizes a field which is responsible for this effect, called the Higgs field symbol: which has the unusual property of a non-zero amplitude in its ground state; i.e., a non-zero vacuum expectation value. It can have this effect because of its unusual "Mexican hat" shaped potential whose lowest "point" is not at its "centre". Below a certain extremely high energy level the existence of this non-zero vacuum expectation spontaneously breaks electroweak gauge symmetry which in turn gives rise to the Higgs mechanism and triggers the acquisition of mass by those particles interacting with the field. This effect occurs because scalar field components of the Higgs field are "absorbed" by the massive bosons as degrees of freedom, and couple to the fermions via Yukawa coupling, thereby producing the expected mass terms. In effect when symmetry breaks under these conditions, the Goldstone bosons that arise *interact* with the Higgs field (and with other particles capable of interacting with the Higgs field) instead of becoming new massless particles, the intractable problems of both underlying theories "neutralise" each other, and the residual outcome is that elementary particles acquire a consistent mass based on how strongly they interact with the Higgs field. It is the simplest known process capable of giving mass to the gauge bosons while remaining compatible with gauge theories. Its quantum would be a scalar boson, known as the Higgs boson.

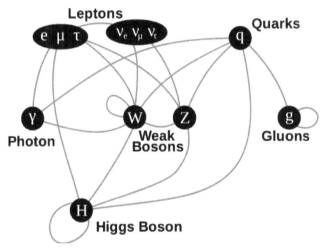

Summary of interactions between certain particles described by the Standard Model.

Properties of the Higgs Field

In the Standard Model, the Higgs field is a scalar tachyonic field – 'scalar' meaning it does not transform under Lorentz transformations, and 'tachyonic' meaning the field (but not the particle) has imaginary mass and in certain configurations must undergo symmetry breaking. It consists of four components, two neutral ones and two charged component fields. Both of the charged components and one of the neutral fields are Goldstone bosons, which act as the longitudinal third-polarization components of the massive W^+, W^-, and Z bosons. The quantum of the remaining neutral component corresponds to (and is theoretically realised as) the massive Higgs boson, this component can interact with fermions via Yukawa coupling to give them mass, as well.

Mathematically, the Higgs field has imaginary mass and is therefore a tachyonic field. While tachyons (particles that move faster than light) are a purely hypothetical concept, fields with imaginary mass have come to play an important role in modern physics. Under no circumstances do any excitations ever propagate faster than light in such theories — the presence or absence of a tachyonic mass has no effect whatsoever on the maximum velocity of signals (there is no violation of causality). Instead of faster-than-light particles, the imaginary mass creates an instability:- any configuration in which one or more field excitations are tachyonic must spontaneously decay, and the resulting configuration contains no physical tachyons. This process is known as tachyon condensation, and is now believed to be the explanation for how the Higgs mechanism itself arises in nature, and therefore the reason behind electroweak symmetry breaking.

Although the notion of imaginary mass might seem troubling, it is only the field, and not the mass itself, that is quantized. Therefore, the field operators at spacelike separated points still commute (or anticommute), and information and particles still do not propagate faster than light. Tachyon condensation drives a physical system that has reached a local limit and might naively be expected to produce physical tachyons, to an alternate stable state where no physical tachyons exist. Once a tachyonic field such as the Higgs field reaches the minimum of the potential, its quanta are not tachyons any more but rather are ordinary particles such as the Higgs boson.

Properties of the Higgs Boson

Since the Higgs field is scalar, the Higgs boson has no spin. The Higgs boson is also its own antiparticle and is CP-even, and has zero electric and colour charge.

The Minimal Standard Model does not predict the mass of the Higgs boson. If that mass is between 115 and 180 GeV/c^2, then the Standard Model can be valid at energy scales all the way up to the Planck scale (10^{19} GeV). Many theorists expect new physics beyond the Standard Model to emerge at the TeV-scale, based on unsatisfactory properties of the Standard Model. The highest possible mass scale allowed for the Higgs boson (or some other electroweak symmetry breaking mechanism) is 1.4 TeV; beyond this point, the Standard Model becomes inconsistent without such a mechanism, because unitarity is violated in certain scattering processes.

It is also possible, although experimentally difficult, to estimate the mass of the Higgs boson indirectly. In the Standard Model, the Higgs boson has a number of indirect effects; most notably, Higgs loops result in tiny corrections to masses of W and Z bosons. Precision measurements of electroweak parameters, such as the Fermi constant and masses of W/Z bosons, can be used to calculate constraints on the mass of the Higgs. As of July 2011, the precision electroweak measurements tell us that the mass of the Higgs boson is likely to be less than about 161 GeV/c^2 at 95% confidence level (this upper limit would increase to 185 GeV/c^2 if the lower bound of 114.4 GeV/c^2 from the LEP-2 direct search is allowed for). These indirect constraints rely on the assumption that the Standard Model is correct. It may still be possible to discover a Higgs boson above these masses if it is accompanied by other particles beyond those predicted by the Standard Model.

Production

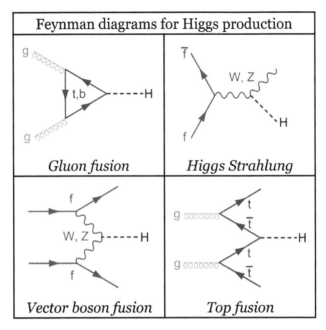

Feynman diagrams for Higgs production	
Gluon fusion	*Higgs Strahlung*
Vector boson fusion	*Top fusion*

If Higgs particle theories are valid, then a Higgs particle can be produced much like other particles that are studied, in a particle collider. This involves accelerating a large number of particles to extremely high energies and extremely close to the speed of light, then allowing them to smash together. Protons and lead ions (the bare nuclei of lead atoms) are used at the LHC. In the extreme energies of these collisions, the desired esoteric particles will occasionally be produced and this can be detected and studied; any absence or difference from theoretical expectations can also be used to improve the theory. The relevant particle theory (in this case the Standard Model) will determine the necessary kinds of collisions and detectors. The Standard Model predicts that Higgs bosons could be formed in a number of ways, although the probability of producing a Higgs boson in any collision is always expected to be very small—for example, only 1 Higgs boson per 10 billion collisions in the Large Hadron Collider.[Note 14] The most common expected processes for Higgs boson production are:

- *Gluon fusion.* If the collided particles are hadrons such as the proton or anti-proton—as is the case in the LHC and Tevatron—then it is most likely that two of the gluons binding the hadron together collide. The easiest way to produce a Higgs particle is if the two gluons combine to form a loop of virtual quarks. Since the coupling of particles to the Higgs boson is proportional to their mass, this process is more likely for heavy particles. In practice it is enough to consider the contributions of virtual top and bottom quarks (the heaviest quarks). This process is the dominant contribution at the LHC and Tevatron being about ten times more likely than any of the other processes.

- *Higgs Strahlung.* If an elementary fermion collides with an anti-fermion—e.g., a quark with an anti-quark or an electron with a positron—the two can merge to form a virtual W or Z boson which, if it carries sufficient energy, can then emit a Higgs boson. This process was the dominant production mode at the LEP, where an electron and a positron collided to form a virtual Z boson, and it was the second largest contribution for Higgs production at the Tevatron. At the LHC this process is only the third largest, because the LHC collides protons with protons, making a quark-antiquark collision less likely than at the Tevatron. Higgs Strahlung is also known as *associated production*.

- *Weak boson fusion.* Another possibility when two (anti-)fermions collide is that the two exchange a virtual W or Z boson, which emits a Higgs boson. The colliding fermions do not need to be the same type. So, for example, an up quark may exchange a Z boson with an anti-down quark. This process is the second most important for the production of Higgs particle at the LHC and LEP.

- *Top fusion.* The final process that is commonly considered is by far the least likely (by two orders of magnitude). This process involves two colliding gluons, which each decay into a heavy quark–antiquark pair. A quark and antiquark from each pair can then combine to form a Higgs particle.

Decay

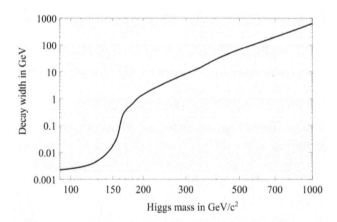

The Standard Model prediction for the decay width of the Higgs particle depends on the value of its mass.

Quantum mechanics predicts that if it is possible for a particle to decay into a set of lighter particles, then it will eventually do so. This is also true for the Higgs boson. The likelihood with which this happens depends on a variety of factors including: the difference in mass, the strength of the interactions, etc. Most of these factors are fixed by the Standard Model, except for the mass of the Higgs boson itself. For a Higgs boson with a mass of 126 GeV/c^2 the SM predicts a mean life time of about 1.6×10^{-22} s.

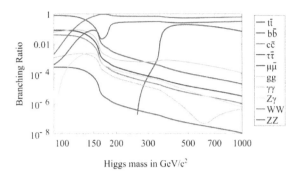

The Standard Model prediction for the branching ratios of the different decay
modes of the Higgs particle depends on the value of its mass.

Since it interacts with all the massive elementary particles of the SM, the Higgs boson
has many different processes through which it can decay. Each of these possible processes has its own probability, expressed as the *branching ratio*; the fraction of the
total number decays that follows that process. The SM predicts these branching ratios
as a function of the Higgs mass.

One way that the Higgs can decay is by splitting into a fermion–antifermion pair. As
general rule, the Higgs is more likely to decay into heavy fermions than light fermions,
because the mass of a fermion is proportional to the strength of its interaction with the
Higgs. By this logic the most common decay should be into a top–antitop quark pair.
However, such a decay is only possible if the Higgs is heavier than ~346 GeV/c^2, twice
the mass of the top quark. For a Higgs mass of 126 GeV/c^2 the SM predicts that the
most common decay is into a bottom–antibottom quark pair, which happens 56.1% of
the time. The second most common fermion decay at that mass is a tau–antitau pair,
which happens only about 6% of the time.

Another possibility is for the Higgs to split into a pair of massive gauge bosons. The
most likely possibility is for the Higgs to decay into a pair of W bosons (the light blue
line in the plot), which happens about 23.1% of the time for a Higgs boson with a mass
of 126 GeV/c^2. The W bosons can subsequently decay either into a quark and an antiquark or into a charged lepton and a neutrino. However, the decays of W bosons into
quarks are difficult to distinguish from the background, and the decays into leptons
cannot be fully reconstructed (because neutrinos are impossible to detect in particle
collision experiments). A cleaner signal is given by decay into a pair of Z-bosons (which
happens about 2.9% of the time for a Higgs with a mass of 126 GeV/c^2), if each of the
bosons subsequently decays into a pair of easy-to-detect charged leptons (electrons or
muons).

Decay into massless gauge bosons (i.e., gluons or photons) is also possible, but requires
intermediate loop of virtual heavy quarks (top or bottom) or massive gauge bosons.
The most common such process is the decay into a pair of gluons through a loop of virtual heavy quarks. This process, which is the reverse of the gluon fusion process men-

tioned above, happens approximately 8.5% of the time for a Higgs boson with a mass of 126 GeV/c^2. Much rarer is the decay into a pair of photons mediated by a loop of W bosons or heavy quarks, which happens only twice for every thousand decays. However, this process is very relevant for experimental searches for the Higgs boson, because the energy and momentum of the photons can be measured very precisely, giving an accurate reconstruction of the mass of the decaying particle.

Alternative Models

The Minimal Standard Model as described above is the simplest known model for the Higgs mechanism with just one Higgs field. However, an extended Higgs sector with additional Higgs particle doublets or triplets is also possible, and many extensions of the Standard Model have this feature. The non-minimal Higgs sector favoured by theory are the two-Higgs-doublet models (2HDM), which predict the existence of a quintet of scalar particles: two CP-even neutral Higgs bosons h° and H°, a CP-odd neutral Higgs boson A°, and two charged Higgs particles $H^\pm$. Supersymmetry ("SUSY") also predicts relations between the Higgs-boson masses and the masses of the gauge bosons, and could accommodate a 125 GeV/c^2 neutral Higgs boson.

The key method to distinguish between these different models involves study of the particles' interactions ("coupling") and exact decay processes ("branching ratios"), which can be measured and tested experimentally in particle collisions. In the Type-I 2HDM model one Higgs doublet couples to up and down quarks, while the second doublet does not couple to quarks. This model has two interesting limits, in which the lightest Higgs couples to just fermions ("gauge-phobic") or just gauge bosons ("fermiophobic"), but not both. In the Type-II 2HDM model, one Higgs doublet only couples to up-type quarks, the other only couples to down-type quarks. The heavily researched Minimal Supersymmetric Standard Model (MSSM) includes a Type-II 2HDM Higgs sector, so it could be disproven by evidence of a Type-I 2HDM Higgs.

In other models the Higgs scalar is a composite particle. For example, in technicolor the role of the Higgs field is played by strongly bound pairs of fermions called techniquarks. Other models, feature pairs of top quarks. In yet other models, there is no Higgs field at all and the electroweak symmetry is broken using extra dimensions.

A one-loop Feynman diagram of the first-order correction to the Higgs mass. In the Standard Model the effects of these corrections are potentially enormous, giving rise to the so-called hierarchy problem.

Further Theoretical Issues and Hierarchy Problem

The Standard Model leaves the mass of the Higgs boson as a parameter to be measured, rather than a value to be calculated. This is seen as theoretically unsatisfactory, particularly as quantum corrections (related to interactions with virtual particles) should apparently cause the Higgs particle to have a mass immensely higher than that observed, but at the same time the Standard Model requires a mass of the order of 100 to 1000 GeV to ensure unitarity (in this case, to unitarise longitudinal vector boson scattering). Reconciling these points appears to require explaining why there is an almost-perfect cancellation resulting in the visible mass of ~ 125 GeV, and it is not clear how to do this. Because the weak force is about 10^{32} times stronger than gravity, and (linked to this) the Higgs boson's mass is so much less than the Planck mass or the grand unification energy, it appears that either there is some underlying connection or reason for these observations which is unknown and not described by the Standard Model, or some unexplained and extremely precise fine-tuning of parameters – however at present neither of these explanations is proven. This is known as a hierarchy problem. More broadly, the hierarchy problem amounts to the worry that a future theory of fundamental particles and interactions should not have excessive fine-tunings or unduly delicate cancellations, and should allow masses of particles such as the Higgs boson to be calculable. The problem is in some ways unique to spin-0 particles (such as the Higgs boson), which can give rise to issues related to quantum corrections that do not affect particles with spin. A number of solutions have been proposed, including supersymmetry, conformal solutions and solutions via extra dimensions such as braneworld models.

There are also issues of quantum triviality, which suggests that it may not be possible to create a consistent quantum field theory involving elementary scalar particles.

Experimental Search

To produce Higgs bosons, two beams of particles are accelerated to very high energies and allowed to collide within a particle detector. Occasionally, although rarely, a Higgs boson will be created fleetingly as part of the collision byproducts. Because the Higgs boson decays very quickly, particle detectors cannot detect it directly. Instead the detectors register all the decay products (the *decay signature*) and from the data the decay process is reconstructed. If the observed decay products match a possible decay process (known as a *decay channel*) of a Higgs boson, this indicates that a Higgs boson may have been created. In practice, many processes may produce similar decay signatures. Fortunately, the Standard Model precisely predicts the likelihood of each of these, and each known process, occurring. So, if the detector detects more decay signatures consistently matching a Higgs boson than would otherwise be expected if Higgs bosons did not exist, then this would be strong evidence that the Higgs boson exists.

Because Higgs boson production in a particle collision is likely to be very rare (1 in 10 billion at the LHC), and many other possible collision events can have similar decay signatures, the data of hundreds of trillions of collisions needs to be analysed and must "show the same picture" before a conclusion about the existence of the Higgs boson can be reached. To conclude that a new particle has been found, particle physicists require that the statistical analysis of two independent particle detectors each indicate that there is lesser than a one-in-a-million chance that the observed decay signatures are due to just background random Standard Model events—i.e., that the observed number of events is more than 5 standard deviations (sigma) different from that expected if there was no new particle. More collision data allows better confirmation of the physical properties of any new particle observed, and allows physicists to decide whether it is indeed a Higgs boson as described by the Standard Model or some other hypothetical new particle.

To find the Higgs boson, a powerful particle accelerator was needed, because Higgs bosons might not be seen in lower-energy experiments. The collider needed to have a high luminosity in order to ensure enough collisions were seen for conclusions to be drawn. Finally, advanced computing facilities were needed to process the vast amount of data (25 petabytes per year as of 2012) produced by the collisions. For the announcement of 4 July 2012, a new collider known as the Large Hadron Collider was constructed at CERN with a planned eventual collision energy of 14 TeV—over seven times any previous collider—and over 300 trillion (3×10^{14}) LHC proton–proton collisions were analysed by the LHC Computing Grid, the world's largest computing grid (as of 2012), comprising over 170 computing facilities in a worldwide network across 36 countries.

Search before 4 July 2012

The first extensive search for the Higgs boson was conducted at the Large Electron–Positron Collider (LEP) at CERN in the 1990s. At the end of its service in 2000, LEP had found no conclusive evidence for the Higgs. This implied that if the Higgs boson were to exist it would have to be heavier than 114.4 GeV/c^2.

The search continued at Fermilab in the United States, where the Tevatron—the collider that discovered the top quark in 1995—had been upgraded for this purpose. There was no guarantee that the Tevatron would be able to find the Higgs, but it was the only supercollider that was operational since the Large Hadron Collider (LHC) was still under construction and the planned Superconducting Super Collider had been cancelled in 1993 and never completed. The Tevatron was only able to exclude further ranges for the Higgs mass, and was shut down on 30 September 2011 because it no longer could keep up with the LHC. The final analysis of the data excluded the possibility of a Higgs boson with a mass between 147 GeV/c^2 and 180 GeV/c^2. In addition, there was a small (but not significant) excess of events possibly indicating a Higgs boson with a mass between 115 GeV/c^2 and 140 GeV/c^2.

The Large Hadron Collider at CERN in Switzerland, was designed specifically to be able to either confirm or exclude the existence of the Higgs boson. Built in a 27 km tunnel under the ground near Geneva originally inhabited by LEP, it was designed to collide two beams of protons, initially at energies of 3.5 TeV per beam (7 TeV total), or almost 3.6 times that of the Tevatron, and upgradeable to 2 × 7 TeV (14 TeV total) in future. Theory suggested if the Higgs boson existed, collisions at these energy levels should be able to reveal it. As one of the most complicated scientific instruments ever built, its operational readiness was delayed for 14 months by a magnet quench event nine days after its inaugural tests, caused by a faulty electrical connection that damaged over 50 superconducting magnets and contaminated the vacuum system.

Data collection at the LHC finally commenced in March 2010. By December 2011 the two main particle detectors at the LHC, ATLAS and CMS, had narrowed down the mass range where the Higgs could exist to around 116-130 GeV (ATLAS) and 115-127 GeV (CMS). There had also already been a number of promising event excesses that had "evaporated" and proven to be nothing but random fluctuations. However, from around May 2011, both experiments had seen among their results, the slow emergence of a small yet consistent excess of gamma and 4-lepton decay signatures and several other particle decays, all hinting at a new particle at a mass around 125 GeV. By around November 2011, the anomalous data at 125 GeV was becoming "too large to ignore" (although still far from conclusive), and the team leaders at both ATLAS and CMS each privately suspected they might have found the Higgs. On November 28, 2011, at an internal meeting of the two team leaders and the director general of CERN, the latest analyses were discussed outside their teams for the first time, suggesting both ATLAS and CMS might be converging on a possible shared result at 125 GeV, and initial preparations commenced in case of a successful finding. While this information was not known publicly at the time, the narrowing of the possible Higgs range to around 115–130 GeV and the repeated observation of small but consistent event excesses across multiple channels at both ATLAS and CMS in the 124-126 GeV region (described as "tantalising hints" of around 2-3 sigma) were public knowledge with "a lot of interest". It was therefore widely anticipated around the end of 2011, that the LHC would provide sufficient data to either exclude or confirm the finding of a Higgs boson by the end of 2012, when their 2012 collision data (with slightly higher 8 TeV collision energy) had been examined.

Discovery of Candidate Boson at CERN

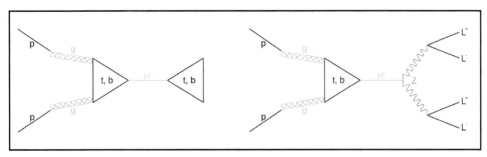

Feynman diagrams showing the cleanest channels associated with the low-mass (~125 GeV) Higgs boson candidate observed by ATLAS and CMS at the LHC. The dominant production mechanism at this mass involves two gluons from each proton fusing to a Top-quark Loop, which couples strongly to the Higgs field to produce a Higgs boson.

Left: Diphoton channel: Boson subsequently decays into 2 gamma ray photons by virtual interaction with a W boson loop or top quark loop.

Right: 4-Lepton "golden channel": Boson emits 2 Z bosons, which each decay into 2 leptons (electrons, muons).

Experimental analysis of these channels reached a significance of more than 5 sigma in both experiments.

On 22 June 2012 CERN announced an upcoming seminar covering tentative findings for 2012, and shortly afterwards (from around 1 July 2012 according to an analysis of the spreading rumour in social media) rumours began to spread in the media that this would include a major announcement, but it was unclear whether this would be a stronger signal or a formal discovery. Speculation escalated to a "fevered" pitch when reports emerged that Peter Higgs, who proposed the particle, was to be attending the seminar, and that "five leading physicists" had been invited – generally believed to signify the five living 1964 authors – with Higgs, Englert, Guralnik, Hagen attending and Kibble confirming his invitation (Brout having died in 2011).

On 4 July 2012 both of the CERN experiments announced they had independently made the same discovery: CMS of a previously unknown boson with mass 125.3 ± 0.6 GeV/c^2 and ATLAS of a boson with mass 126.0 ± 0.6 GeV/c^2. Using the combined analysis of two interaction types (known as 'channels'), both experiments independently reached a local significance of 5 sigma — implying that the probability of getting at least as strong a result by chance alone is less than 1 in 3 million. When additional channels were taken into account, the CMS significance was reduced to 4.9 sigma.

The two teams had been working 'blinded' from each other from around late 2011 or early 2012, meaning they did not discuss their results with each other, providing additional certainty that any common finding was genuine validation of a particle. This level of evidence, confirmed independently by two separate teams and experiments, meets the formal level of proof required to announce a confirmed discovery.

On 31 July 2012, the ATLAS collaboration presented additional data analysis on the "observation of a new particle", including data from a third channel, which improved the significance to 5.9 sigma (1 in 588 million chance of obtaining at least as strong evidence by random background effects alone) and mass 126.0 ± 0.4 (stat) ± 0.4 (sys) GeV/c^2, and CMS improved the significance to 5-sigma and mass 125.3 ± 0.4 (stat) ± 0.5 (sys) GeV/c^2.

The New Particle Tested as a Possible Higgs Boson

Following the 2012 discovery, it was still unconfirmed whether or not the 125 GeV/c^2 particle was a Higgs boson. On one hand, observations remained consistent with the observed particle being the Standard Model Higgs boson, and the particle decayed into at least some of the predicted channels. Moreover, the production rates and branching ratios for the observed channels broadly matched the predictions by the Standard Model within the experimental uncertainties. However, the experimental uncertainties currently still left room for alternative explanations, meaning an announcement of the discovery of a Higgs boson would have been premature. To allow more opportunity for data collection, the LHC's proposed 2012 shutdown and 2013–14 upgrade were postponed by 7 weeks into 2013.

In November 2012, in a conference in Kyoto researchers said evidence gathered since July was falling into line with the basic Standard Model more than its alternatives, with a range of results for several interactions matching that theory's predictions. Physicist Matt Strassler highlighted "considerable" evidence that the new particle is not a pseudoscalar negative parity particle (consistent with this required finding for a Higgs boson), "evaporation" or lack of increased significance for previous hints of non-Standard Model findings, expected Standard Model interactions with W and Z bosons, absence of "significant new implications" for or against supersymmetry, and in general no significant deviations to date from the results expected of a Standard Model Higgs boson. However some kinds of extensions to the Standard Model would also show very similar results; so commentators noted that based on other particles that are still being understood long after their discovery, it may take years to be sure, and decades to fully understand the particle that has been found.

These findings meant that as of January 2013, scientists were very sure they had found an unknown particle of mass ~ 125 GeV/c^2, and had not been misled by experimental error or a chance result. They were also sure, from initial observations, that the new particle was some kind of boson. The behaviours and properties of the particle, so far as examined since July 2012, also seemed quite close to the behaviours expected of a Higgs boson. Even so, it could still have been a Higgs boson or some other unknown boson, since future tests could show behaviours that do not match a Higgs boson, so as of December 2012 CERN still only stated that the new particle was "consistent with" the Higgs boson, and scientists did not yet positively say it was the Higgs boson. Despite this, in late 2012, widespread media reports announced (incorrectly) that a Higgs boson had been confirmed during the year.

In January 2013, CERN director-general Rolf-Dieter Heuer stated that based on data analysis to date, an answer could be possible 'towards' mid-2013, and the deputy chair of physics at Brookhaven National Laboratory stated in February 2013 that a "definitive" answer might require "another few years" after the collider's 2015 restart. In early March 2013, CERN Research Director Sergio Bertolucci stated that confirming spin-

0 was the major remaining requirement to determine whether the particle is at least some kind of Higgs boson.

Preliminary Confirmation of Existence and Current Status

On 14 March 2013 CERN confirmed that:

> "CMS and ATLAS have compared a number of options for the spin-parity of this particle, and these all prefer no spin and even parity [two fundamental criteria of a Higgs boson consistent with the Standard Model]. This, coupled with the measured interactions of the new particle with other particles, strongly indicates that it is a Higgs boson."

This also makes the particle the first elementary scalar particle to be discovered in nature.

Examples of tests used to validate whether the 125 GeV particle is a Higgs boson:

Requirement	How tested / explanation	Current status (March 2013)
Zero spin	Examining decay patterns. Spin-1 had been ruled out at the time of initial discovery by the observed decay to two photons (γγ), leaving spin-0 and spin-2 as remaining candidates.	Spin-0 tentatively confirmed. The spin-2 hypothesis is excluded with a confidence level exceeding 99.9%.
$^+$ and not $^-$ parity	Studying the angles at which decay products fly apart. Negative parity was also disfavoured if spin-0 was confirmed.	Even parity tentatively confirmed. The spin-0 negative parity hypothesis is excluded with a confidence level exceeding 99.9%.
Decay channels (outcomes of particle decaying) are as predicted	The Standard Model predicts the decay patterns of a 125–126 GeV Higgs boson. Are these all being seen, and at the right rates? Particularly significant, we should observe decays into pairs of photons (γγ), W and Z bosons (WW and ZZ), bottom quarks (bb), and tau leptons (ττ), among the possible outcomes.	γγ, ττ, WW and ZZ observed; bb not yet confirmed. Some branching levels (decay rates) are a little higher than expected in preliminary results, in particular H → γγ, which gives a peak at ATLAS a little higher than that seen in 4-lepton decays and at CMS.
Couples to mass (i.e., strength of interaction with Standard Model particles proportional to their mass)	Particle physicist Adam Falkowski states that the essential qualities of a Higgs boson are that it is a spin-0 (scalar) particle which *also* couples to mass (W and Z bosons); proving spin-0 alone is insufficient.	Couplings to mass strongly evidenced ("At 95% confidence level c_V is within 15% of the standard model value $c_V=1$").
Higher energy results remain consistent	After the LHC's 2015 restart at the LHC's full planned energies of 13–14 TeV, searches for multiple Higgs particles (as predicted in some theories) and tests targeting other versions of particle theory will take place. These higher energy results must continue to give results consistent with Higgs theories	To be studied following LHC upgrade

Names Used by Physicists

The name most strongly associated with the particle and field is the Higgs boson and Higgs field. For some time the particle was known by a combination of its PRL author names (including at times Anderson), for example the Brout–Englert–Higgs particle, the Anderson-Higgs particle, or the Englert–Brout–Higgs–Guralnik–Hagen–Kibble mechanism, and these are still used at times. Fueled in part by the issue of recognition and a potential shared Nobel Prize, the most appropriate name is still occasionally a topic of debate as of 2012. (Higgs himself prefers to call the particle either by an acronym of all those involved, or "the scalar boson", or "the so-called Higgs particle".)

A considerable amount has been written on how Higgs' name came to be exclusively used. Two main explanations are offered.

Reason	Basis of explanation
Higgs undertook a step which was either unique, clearer or more explicit in his paper in formally predicting and examining the particle.	Of the PRL papers' authors, only the paper by Higgs *explicitly* offered as a prediction that a massive particle would exist and calculated some of its properties; he was therefore "the first to postulate the existence of a massive particle" according to *Nature*. Physicist and author Frank Close and physicist-blogger Peter Woit both comment that the paper by GHK was also completed after Higgs and Brout–Englert were submitted to Physical Review Letters. and that Higgs alone had drawn attention to a predicted massive *scalar* boson, while all others had focused on the massive *vector* bosons; In this way, Higgs' contribution also provided experimentalists with a crucial "concrete target" needed to test the theory. However, in Higgs' view, Brout and Englert did not explicitly mention the boson since its existence is plainly obvious in their work, while according to Guralnik the GHK paper was a complete analysis of the entire symmetry breaking mechanism whose mathematical rigour is absent from the other two papers, and a massive particle may exist in some solutions. Higgs' paper also provided an "especially sharp" statement of the challenge and its solution according to science historian David Kaiser.
The name was popularised in the 1970s due to its use as a convenient shorthand or because of a mistake in citing.	Many accounts (including Higgs' own) credit the "Higgs" name to physicist Benjamin Lee (in Korean: Lee Whi-soh). Lee was a significant populist for the theory in its early stages, and habitually attached the name "Higgs" as a "convenient shorthand" for its components from 1972 and in at least one instance from as early as 1966. Although Lee clarified in his footnotes that "'Higgs' is an abbreviation for Higgs, Kibble, Guralnik, Hagen, Brout, Englert", his use of the term (and perhaps also Steven Weinberg's mistaken cite of Higgs' paper as the first in his seminal 1967 paper) meant that by around 1975–76 others had also begun to use the name 'Higgs' exclusively as a shorthand.

The Higgs boson is often referred to as the "God particle" in popular media outside the scientific community. The nickname comes from the title of the 1993 book on the Higgs boson and particle physics, *The God Particle: If the Universe Is the Answer, What Is the Question?* by Nobel Physics prizewinner and Fermilab director Leon Lederman. Lederman wrote it in the context of failing US government support for the Superconducting Super Collider, a part-constructed titanic competitor to the Large Hadron Collider with planned collision energies of 2×20 TeV that was championed by Lederman since its 1983 inception and shut down in 1993. The book sought in part to promote awareness of the significance and need for such a project in the face of its possible loss of funding. Lederman, a leading researcher in the field, wanted to title his book *The Goddamn Particle: If the Universe is the Answer, What is the Question?* But his editor decided that the title was too controversial and convinced Lederman to change the title to *The God Particle: If the Universe is the Answer, What is the Question?*

While media use of this term may have contributed to wider awareness and interest, many scientists feel the name is inappropriate since it is sensational hyperbole and misleads readers; the particle also has nothing to do with God, leaves open numerous questions in fundamental physics, and does not explain the ultimate origin of the universe. Higgs, an atheist, was reported to be displeased and stated in a 2008 interview that he found it "embarrassing" because it was "the kind of misuse... which I think might offend some people". Science writer Ian Sample stated in his 2010 book on the search that the nickname is "universally hate[d]" by physicists and perhaps the "worst derided" in the history of physics, but that (according to Lederman) the publisher rejected all titles mentioning "Higgs" as unimaginative and too unknown.

Lederman begins with a review of the long human search for knowledge, and explains that his tongue-in-cheek title draws an analogy between the impact of the Higgs field on the fundamental symmetries at the Big Bang, and the apparent chaos of structures, particles, forces and interactions that resulted and shaped our present universe, with the biblical story of Babel in which the primordial single language of early Genesis was fragmented into many disparate languages and cultures.

Today ... we have the standard model, which reduces all of reality to a dozen or so particles and four forces. ... It's a hard-won simplicity [...and...] remarkably accurate. But it is also incomplete and, in fact, internally inconsistent... This boson is so central to the state of physics today, so crucial to our final understanding of the structure of matter, yet so elusive, that I have given it a nickname: the God Particle. Why God Particle? Two reasons. One, the publisher wouldn't let us call it the Goddamn Particle, though that might be a more appropriate title, given its villainous nature and the expense it is causing. And two, there is a connection, of sorts, to another book, a *much* older one...

— *Leon M. Lederman and Dick Teresi, The God Particle: If the Universe is the Answer, What is the Question p. 22*

Lederman asks whether the Higgs boson was added just to perplex and confound those seeking knowledge of the universe, and whether physicists will be confounded by it as recounted in that story, or ultimately surmount the challenge and understand "how beautiful is the universe [God has] made".

Other Proposals

A renaming competition by British newspaper *The Guardian* in 2009 resulted in their science correspondent choosing the name "the champagne bottle boson" as the best submission: "The bottom of a champagne bottle is in the shape of the Higgs potential and is often used as an illustration in physics lectures. So it's not an embarrassingly grandiose name, it is memorable, and [it] has some physics connection too." The name *Higgson* was suggested as well, in an opinion piece in the Institute of Physics' online publication *physicsworld.com*.

Media Explanations and Analogies

There has been considerable public discussion of analogies and explanations for the Higgs particle and how the field creates mass, including coverage of explanatory attempts in their own right and a competition in 1993 for the best popular explanation by then-UK Minister for Science Sir William Waldegrave and articles in newspapers worldwide.

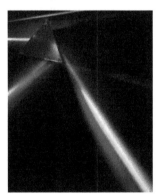

Photograph of light passing through a dispersive prism: the rainbow effect arises because photons are not all affected to the same degree by the dispersive material of the prism.

An educational collaboration involving an LHC physicist and a High School Teachers at CERN educator suggests that dispersion of light – responsible for the rainbow and dispersive prism – is a useful analogy for the Higgs field's symmetry breaking and mass-causing effect.

Symmetry breaking in optics	In a vacuum, light of all colours (or photons of all wavelengths) travels at the same velocity, a symmetrical situation. In some substances such as glass, water or air, this symmetry is broken. The result is that light of different wavelengths appears to have different velocities (as seen from outside).
Symmetry breaking in particle physics	In 'naive' gauge theories, gauge bosons and other fundamental particles are all massless – also a symmetrical situation. In the presence of the Higgs field this symmetry is broken. The result is that particles of different types will have different masses.

Matt Strassler uses electric fields as an analogy:

Some particles interact with the Higgs field while others don't. Those particles that feel the Higgs field act as if they have mass. Something similar happens in an electric field – charged objects are pulled around and neutral objects can sail through unaffected. So you can think of the Higgs search as an attempt to make waves in the Higgs field *[create Higgs bosons]* to prove it's really there.

A similar explanation was offered by *The Guardian*:

The Higgs boson is essentially a ripple in a field said to have emerged at the birth of the universe and to span the cosmos to this day ... The particle is crucial however: it is the smoking gun, the evidence required to show the theory is right.

The Higgs field's effect on particles was famously described by physicist David Miller as akin to a room full of political party workers spread evenly throughout a room: the crowd gravitates to and slows down famous people but does not slow down others. He also drew attention to well-known effects in solid state physics where an electron's effective mass can be much greater than usual in the presence of a crystal lattice.

Analogies based on drag effects, including analogies of "syrup" or "molasses" are also well known, but can be somewhat misleading since they may be understood (incorrectly) as saying that the Higgs field simply resists some particles' motion but not others' – a simple resistive effect could also conflict with Newton's third law.

Recognition and Awards

There has been considerable discussion of how to allocate the credit if the Higgs boson is proven, made more pointed as a Nobel prize had been expected, and the very wide basis of people entitled to consideration. These include a range of theoreticians who made the Higgs mechanism theory possible, the theoreticians of the 1964 PRL papers (including Higgs himself), the theoreticians who derived from these, a working electroweak theory and the Standard Model itself, and also the experimentalists at CERN and other institutions who made possible the proof of the Higgs field and boson in reality. The Nobel prize has a limit of 3 persons to share an award, and some possible winners are already prize holders for other work, or are deceased (the prize is only awarded to persons in their lifetime). Existing prizes for works relating to the Higgs field, boson, or mechanism include:

- Nobel Prize in Physics (1979) – Glashow, Salam, and Weinberg, *for contributions to the theory of the unified weak and electromagnetic interaction between elementary particles*

- Nobel Prize in Physics (1999) – 't Hooft and Veltman, *for elucidating the quantum structure of electroweak interactions in physics*

- Nobel Prize in Physics (2008) – Nambu (shared), *for the discovery of the mechanism of spontaneous broken symmetry in subatomic physics*

- J. J. Sakurai Prize for Theoretical Particle Physics (2010) – Hagen, Englert, Guralnik, Higgs, Brout, and Kibble, *for elucidation of the properties of spontaneous symmetry breaking in four-dimensional relativistic gauge theory and of the mechanism for the consistent generation of vector boson masses* (for the 1964 papers described above)

- Wolf Prize (2004) – Englert, Brout, and Higgs

- Nobel Prize in Physics (2013) - Peter Higgs and François Englert, *for the theoretical discovery of a mechanism that contributes to our understanding of the origin of mass of subatomic particles, and which recently was confirmed through the discovery of the predicted fundamental particle, by the ATLAS and CMS experiments at CERN's Large Hadron Collider*

Additionally Physical Review Letters' 50-year review (2008) recognized the 1964 PRL symmetry breaking papers and Weinberg's 1967 paper *A model of Leptons* (the most cited paper in particle physics, as of 2012) "milestone Letters".

Following reported observation of the Higgs-like particle in July 2012, several Indian media outlets reported on the supposed neglect of credit to Indian physicist Satyendra Nath Bose after whose work in the 1920s the class of particles "bosons" is named (although physicists have described Bose's connection to the discovery as tenuous).

Technical Aspects and Mathematical Formulation

In the Standard Model, the Higgs field is a four-component scalar field that forms a complex doublet of the weak isospin SU(2) symmetry:

$$\phi = \frac{1}{\sqrt{2}} \begin{pmatrix} \phi^1 + i\phi^2 \\ \phi^0 + i\phi^3 \end{pmatrix}, \tag{1}$$

while the field has charge +1/2 under the weak hypercharge U(1) symmetry (in the convention where the electric charge, Q, the weak isospin, I_3, and the weak hypercharge, Y, are related by $Q = I_3 + Y$).

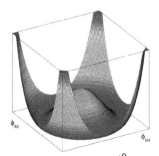

The potential for the Higgs field, plotted as function of ϕ^0 and ϕ^3. It has a *Mexican-hat* or *champagne-bottle profile* at the ground.

The Higgs part of the Lagrangian is

$$\mathcal{L}_H = \left\| \left(\partial_\mu - igW_\mu^a \tau^a - i\frac{g'}{2}B_\mu \right)\phi \right\|^2 + \mu^2 \phi^\dagger \phi - \lambda(\phi^\dagger \phi)^2, \tag{2}$$

where W_μ^a and B_μ are the gauge bosons of the SU(2) and U(1) symmetries, g and g' their respective coupling constants, $\tau^a = \sigma^a / 2$ (where σ^a are the Pauli matrices) a complete set generators of the SU(2) symmetry, and $\lambda > 0$ and $\mu^2 > 0$, so that the ground state breaks the SU(2) symmetry. The ground state of the Higgs field (the bottom of the potential) is degenerate with different ground states related to each other by a SU(2) gauge transformation. It is always possible to pick a gauge such that in the ground state $\phi^1 = \phi^2 = \phi^3 = 0.$. The expectation value of ϕ^0 in the ground state (the vacuum expectation value or vev) is then $\left\langle \phi^0 \right\rangle = \frac{v}{\sqrt{2}}$, where $v = \frac{|\mu|}{\sqrt{\lambda}}$. The mea-sured value of this parameter is ~246 GeV/c². It has units of mass, and is the only free parameter of the Standard Model that is not a dimensionless number. Quadratic terms in W_μ and B_μ arise, which give masses to the W and Z bosons:

$$M_W = \frac{v|g|}{2}, \tag{3}$$

$$M_Z = \frac{v\sqrt{g^2 + g'^2}}{2}, \tag{4}$$

with their ratio determining the Weinberg angle, $\cos\theta_W = \dfrac{M_W}{M_Z} = \dfrac{|g|}{\sqrt{g^2 + g'^2}}$, and leave a massless U(1) photon, γ .

The quarks and the leptons interact with the Higgs field through Yukawa interaction terms:

$$\mathcal{L}_Y = \begin{aligned} &-\lambda_u^{ij} \frac{\phi^0 - i\phi^3}{\sqrt{2}} \bar{u}_L^i u_R^j + \lambda_u^{ij} \frac{\phi^1 - i\phi^2}{\sqrt{2}} \bar{d}_L^i u_R^j \\ &-\lambda_d^{ij} \frac{\phi^0 + i\phi^3}{\sqrt{2}} \bar{d}_L^i d_R^j - \lambda_d^{ij} \frac{\phi^1 + i\phi^2}{\sqrt{2}} \bar{u}_L^i d_R^j \\ &-\lambda_e^{ij} \frac{\phi^0 + i\phi^3}{\sqrt{2}} \bar{e}_L^i e_R^j - \lambda_e^{ij} \frac{\phi^1 + i\phi^2}{\sqrt{2}} \bar{v}_L^i e_R^j + \text{h.c.}, \end{aligned} \tag{5}$$

where $(d, u, e, v)_{L,R}^i$ are left-handed and right-handed quarks and leptons of the ith generation, $\lambda_{u,d,e}^{ij}$ are matrices of Yukawa couplings where h.c. denotes the hermitian conjugate terms. In the symmetry breaking ground state, only the terms containing ϕ^0 remain, giving rise to mass terms for the fermions. Rotating the quark and lepton fields to the basis where the matrices of Yukawa couplings are diagonal, one gets

$$\mathcal{L}_m = -m_u^i \bar{u}_L^i u_R^i - m_d^i \bar{d}_L^i d_R^i - m_e^i \bar{e}_L^i e_R^i + \text{h.c.}, \tag{6}$$

where the masses of the fermions are $m_{u,d,e}^i = \lambda_{u,d,e}^i v / \sqrt{2}$, and $\lambda_{u,d,e}^i$ denote the eigenvalues of the Yukawa matrices.

Fermion

In particle physics, a fermion (a name coined by Paul Dirac from the surname of Enrico Fermi) is any particle characterized by Fermi–Dirac statistics. These particles obey the Pauli exclusion principle. Fermions include all quarks and leptons, as well as any composite particle made of an odd number of these, such as all baryons and many atoms and nuclei. Fermions differ from bosons, which obey Bose–Einstein statistics.

Enrico Fermi

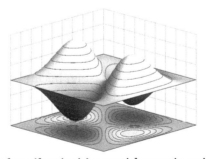

Antisymmetric wavefunction for a (fermionic) 2-particle state in an infinite square well potential.

A fermion can be an elementary particle, such as the electron, or it can be a composite particle, such as the proton. According to the spin-statistics theorem in any reasonable relativistic quantum field theory, particles with integer spin are bosons, while particles with half-integer spin are fermions.

Besides this spin characteristic, fermions have another specific property: they possess conserved baryon or lepton quantum numbers. Therefore, what is usually referred as the spin statistics relation is in fact a spin statistics-quantum number relation.

As a consequence of the Pauli exclusion principle, only one fermion can occupy a particular quantum state at any given time. If multiple fermions have the same spatial probability distribution, then at least one property of each fermion, such as its spin, must be different. Fermions are usually associated with matter, whereas bosons are generally force carrier particles, although in the current state of particle physics the distinction between the two concepts is unclear. Weakly interacting fermions can also display bosonic behavior under extreme conditions. At low temperature fermions show superfluidity for uncharged particles and superconductivity for charged particles.

Composite fermions, such as protons and neutrons, are the key building blocks of everyday matter.

Elementary Fermions

The Standard Model recognizes two types of elementary fermions: quarks and leptons. In all, the model distinguishes 24 different fermions. There are six quarks (up, down, strange, charm, bottom and top quarks), and six leptons (electron, electron neutrino, muon, muon neutrino, tau particle and tau neutrino), along with the corresponding antiparticle of each of these.

Mathematically, fermions come in three types - Weyl fermions (massless), Dirac fermions (massive), and Majorana fermions (each its own antiparticle). Most Standard Model fermions are believed to be Dirac fermions, although it is unknown at this time whether the neutrinos are Dirac or Majorana fermions. Dirac fermions can be treated as a combination of two Weyl fermions. In July 2015, Weyl fermions have been experimentally realized in Weyl semimetals.

Composite Fermions

Composite particles (such as hadrons, nuclei, and atoms) can be bosons or fermions depending on their constituents. More precisely, because of the relation between spin and statistics, a particle containing an odd number of fermions is itself a fermion. It will have half-integer spin.

Examples include the following:

*   A baryon, such as the proton or neutron, contains three fermionic quarks and thus it is a fermion.

*   The nucleus of a carbon-13 atom contains six protons and seven neutrons and is therefore a fermion.

*   The atom helium-3 (^{3}He) is made of two protons, one neutron, and two electrons, and therefore it is a fermion.

The number of bosons within a composite particle made up of simple particles bound with a potential has no effect on whether it is a boson or a fermion.

Fermionic or bosonic behavior of a composite particle (or system) is only seen at large (compared to size of the system) distances. At proximity, where spatial structure begins to be important, a composite particle (or system) behaves according to its constituent makeup.

Fermions can exhibit bosonic behavior when they become loosely bound in pairs. This is the origin of superconductivity and the superfluidity of helium-3: in superconducting materials, electrons interact through the exchange of phonons, forming Cooper pairs, while in helium-3, Cooper pairs are formed via spin fluctuations.

The quasiparticles of the fractional quantum Hall effect are also known as composite fermions, which are electrons with an even number of quantized vortices attached to them.

Skyrmions

In a quantum field theory, there can be field configurations of bosons which are topologically twisted. These are coherent states (or solitons) which behave like a particle, and they can be fermionic even if all the constituent particles are bosons. This was discovered by Tony Skyrme in the early 1960s, so fermions made of bosons are named skyrmions after him.

Skyrme's original example involved fields which take values on a three-dimensional sphere, the original nonlinear sigma model which describes the large distance behavior of pions. In Skyrme's model, reproduced in the large N or string approximation to quantum chromodynamics (QCD), the proton and neutron are fermionic topological solitons of the pion field.

Whereas Skyrme's example involved pion physics, there is a much more familiar example in quantum electrodynamics with a magnetic monopole. A bosonic monopole with the smallest possible magnetic charge and a bosonic version of the electron will form a fermionic dyon.

The analogy between the Skyrme field and the Higgs field of the electroweak sector has been used to postulate that all fermions are skyrmions. This could explain why all known fermions have baryon or lepton quantum numbers and provide a physical mechanism for the Pauli exclusion principle.

Antimatter

In particle physics, antimatter is a material composed of antiparticles, which have the same

mass as particles of ordinary matter but opposite charges, as well as other particle properties such as lepton and baryon numbers. Collisions between particles and antiparticles lead
to the annihilation of both, giving rise to variable proportions of intense photons (gamma
rays), neutrinos, and less massive particle–antiparticle pairs. The total consequence of annihilation is a release of energy available for work, proportional to the total matter and
antimatter mass, in accord with the mass–energy equivalence equation, $E = mc^2$.

Antiparticles bind with each other to form antimatter, just as ordinary particles bind
to form normal matter. For example, a positron (the antiparticle of the electron) and
an antiproton (the antiparticle of the proton) can form an antihydrogen atom. Physical
principles indicate that complex antimatter atomic nuclei are possible, as well as anti-atoms corresponding to the known chemical elements. Studies of cosmic rays have
identified both positrons and antiprotons, presumably produced by collisions between
particles of ordinary matter. Satellite-based searches of cosmic rays for antideuteron
and antihelium particles have yielded nothing.

There is considerable speculation as to why the observable universe is composed almost entirely of ordinary matter, as opposed to an even mixture of matter and antimatter. This asymmetry of matter and antimatter in the visible universe is one of the great
unsolved problems in physics. The process by which this inequality between particles
and antiparticles developed is called baryogenesis.

Antimatter in the form of anti-atoms is one of the most difficult materials to produce.
Antimatter in the form of individual anti-particles, however, is commonly produced by
particle accelerators and in some types of radioactive decay. The nuclei of antihelium
(both helium-3 and helium-4) have been artificially produced with difficulty. These are
the most complex anti-nuclei so far observed.

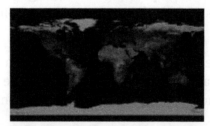

There are some 500 terrestrial gamma-ray flashes daily. The red dots show those
spotted by the Fermi Gamma-ray Space Telescope in 2010.

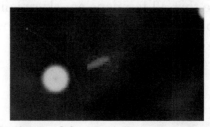

A video showing how scientists used the Fermi Gamma-ray Space Telescope's gamma-ray
detector to uncover bursts of antimatter from thunderstorms

History of the Concept

The idea of negative matter appears in past theories of matter that have now been abandoned. Using the once popular vortex theory of gravity, the possibility of matter with negative gravity was discussed by William Hicks in the 1880s. Between the 1880s and the 1890s, Karl Pearson proposed the existence of "squirts" and sinks of the flow of aether. The squirts represented normal matter and the sinks represented negative matter. Pearson's theory required a fourth dimension for the aether to flow from and into.

The term antimatter was first used by Arthur Schuster in two rather whimsical letters to *Nature* in 1898, in which he coined the term. He hypothesized antiatoms, as well as whole antimatter solar systems, and discussed the possibility of matter and antimatter annihilating each other. Schuster's ideas were not a serious theoretical proposal, merely speculation, and like the previous ideas, differed from the modern concept of antimatter in that it possessed negative gravity.

The modern theory of antimatter began in 1928, with a paper by Paul Dirac. Dirac realised that his relativistic version of the Schrödinger wave equation for electrons predicted the possibility of antielectrons. These were discovered by Carl D. Anderson in 1932 and named positrons (a portmanteau of "positive electron"). Although Dirac did not himself use the term antimatter, its use follows on naturally enough from antielectrons, antiprotons, etc. A complete periodic table of antimatter was envisaged by Charles Janet in 1929.

The Feynman–Stueckelberg interpretation states that antimatter and antiparticles are regular particles traveling backward in time.

Notation

One way to denote an antiparticle is by adding a bar over the particle's symbol. For example, the proton and antiproton are denoted as p and p, respectively. The same rule applies if one were to address a particle by its constituent components. A proton is made up of uud quarks, so an antiproton must therefore be formed from uud antiquarks. Another convention is to distinguish particles by their electric charge. Thus, the electron and positron are denoted simply as e– and e+ respectively. However, to prevent confusion, the two conventions are never mixed.

Origin and Asymmetry

Almost all matter observable from the Earth seems to be made of matter rather than antimatter. If antimatter-dominated regions of space existed, the gamma rays produced in annihilation reactions along the boundary between matter and antimatter regions would be detectable.

Antiparticles are created everywhere in the universe where high-energy particle collisions take place. High-energy cosmic rays impacting Earth's atmosphere (or any other matter in the Solar System) produce minute quantities of antiparticles in the resulting particle jets, which are immediately annihilated by contact with nearby matter. They may similarly be produced in regions like the center of the Milky Way and other galaxies, where very energetic celestial events occur (principally the interaction of relativistic jets with the interstellar medium). The presence of the resulting antimatter is detectable by the two gamma rays produced every time positrons annihilate with nearby matter. The frequency and wavelength of the gamma rays indicate that each carries 511 keV of energy (i.e., the rest mass of an electron multiplied by c^2).

Recent observations by the European Space Agency's INTEGRAL satellite may explain the origin of a giant antimatter cloud surrounding the galactic center. The observations show that the cloud is asymmetrical and matches the pattern of X-ray binaries (binary star systems containing black holes or neutron stars), mostly on one side of the galactic center. While the mechanism is not fully understood, it is likely to involve the production of electron–positron pairs, as ordinary matter gains kinetic energy while falling into a stellar remnant.

Antimatter may exist in relatively large amounts in far-away galaxies due to cosmic inflation in the primordial time of the universe. Antimatter galaxies, if they exist, are expected to have the same chemistry and absorption and emission spectra as normal-matter galaxies, and their astronomical objects would be observationally identical, making them difficult to distinguish. NASA is trying to determine if such galaxies exist by looking for X-ray and gamma-ray signatures of annihilation events in colliding superclusters.

Natural Production

Positrons are produced naturally in β^+ decays of naturally occurring radioactive isotopes (for example, potassium-40) and in interactions of gamma quanta (emitted by radioactive nuclei) with matter. Antineutrinos are another kind of antiparticle created by natural radioactivity (β^- decay). Many different kinds of antiparticles are also produced by (and contained in) cosmic rays. In January 2011, research by the American Astronomical Society discovered antimatter (positrons) originating above thunderstorm clouds; positrons are produced in gamma-ray flashes created by electrons accelerated by strong electric fields in the clouds. Antiprotons have also been found to exist in the Van Allen Belts around the Earth by the PAMELA module.

Antiparticles are also produced in any environment with a sufficiently high temperature (mean particle energy greater than the pair production threshold). During the period of baryogenesis, when the universe was extremely hot and dense, matter and antimatter were continually produced and annihilated. The presence of remaining matter, and absence of detectable remaining antimatter, also called baryon asymmetry, is attributed to CP-violation: a violation of the CP-symmetry relating matter to antimatter. The exact mechanism of this violation during baryogenesis remains a mystery.

Recent observations indicate black holes and neutron stars produce vast amounts of positron-electron plasma via the jets. A possible process is: proton → positron + 938MeV.

Observation in Cosmic Rays

Satellite experiments have found evidence of positrons and a few antiprotons in primary cosmic rays, amounting to less than 1% of the particles in primary cosmic rays. This antimatter cannot all have been created in the Big Bang, but is instead attributed to have been produced by cyclic processes at high energies. For instance, electron-positron pairs may be formed in pulsars, as a magnetized neutron star rotation cycle shears electron-positron pairs from the star surface. Therein the antimatter forms a wind which crashes upon the ejecta of the progenitor supernovae. This weathering takes place as "the cold, magnetized relativistic wind launched by the star hits the non-relativistically expanding ejecta, a shock wave system forms in the impact: the outer one propagates in the ejecta, while a reverse shock propagates back towards the star." The former ejection of matter in the outer shock wave and the latter production of antimatter in the reverse shock wave are steps in a space weather cycle.

Preliminary results from the presently operating Alpha Magnetic Spectrometer (*AMS-02*) on board the International Space Station show that positrons in the cosmic rays arrive with no directionality, and with energies that range from 10 GeV to 250 GeV. In September, 2014, new results with almost twice as much data were presented in a talk at CERN and published in Physical Review Letters. A new measurement of positron fraction up to 500 GeV was reported, showing that positron fraction peaks at a maximum of about 16% of total electron+positron events, around an energy of 275 ± 32 GeV. At higher energies, up to 500 GeV, the ratio of positrons to electrons begins to fall again. The absolute flux of positrons also begins to fall before 500 GeV, but peaks at energies far higher than electron energies, which peak about 10 GeV. These results on interpretation have been suggested to be due to positron production in annihilation events of massive dark matter particles.

Cosmic ray antiprotons also have a much higher energy than their normal-matter counterparts (protons). They arrive at Earth with a characteristic energy maximum of 2 GeV, indicating their production in a fundamentally different process from cosmic ray protons, which on average have only one-sixth of the energy.

There is no evidence of complex antimatter atomic nuclei, such as antihelium nuclei (i.e., anti-alpha particles), in cosmic rays. These are actively being searched for. A prototype of the *AMS-02* designated *AMS-01*, was flown into space aboard the Space Shuttle *Discovery* on STS-91 in June 1998. By not detecting any antihelium at all, the *AMS-01* established an upper limit of 1.1×10^{-6} for the antihelium to helium flux ratio.

Artificial Production

Positrons

Positrons were reported in November 2008 to have been generated by Lawrence Livermore National Laboratory in larger numbers than by any previous synthetic process. A laser drove electrons through a millimeter-radius gold target's nuclei, which caused the incoming electrons to emit energy quanta that decayed into both matter and antimatter. Positrons were detected at a higher rate and in greater density than ever previously detected in a laboratory. Previous experiments made smaller quantities of positrons using lasers and paper-thin targets; however, new simulations showed that short, ultra-intense lasers and millimeter-thick gold are a far more effective source.

Antiprotons, Antineutrons, and Antinuclei

The existence of the antiproton was experimentally confirmed in 1955 by University of California, Berkeley physicists Emilio Segrè and Owen Chamberlain, for which they were awarded the 1959 Nobel Prize in Physics. An antiproton consists of two up antiquarks and one down antiquark (uud). The properties of the antiproton that have been measured all match the corresponding properties of the proton, with the exception of the antiproton having opposite electric charge and magnetic moment from the proton. Shortly afterwards, in 1956, the antineutron was discovered in proton–proton collisions at the Bevatron (Lawrence Berkeley National Laboratory) by Bruce Cork and colleagues.

In addition to antibaryons, anti-nuclei consisting of multiple bound antiprotons and antineutrons have been created. These are typically produced at energies far too high to form antimatter atoms (with bound positrons in place of electrons). In 1965, a group of researchers led by Antonino Zichichi reported production of nuclei of antideuterium at the Proton Synchrotron at CERN. At roughly the same time, observations of antideuterium nuclei were reported by a group of American physicists at the Alternating Gradient Synchrotron at Brookhaven National Laboratory.

Antihydrogen Atoms

In 1995, CERN announced that it had successfully brought into existence nine hot antihydrogen atoms by implementing the SLAC/Fermilab concept during the PS210 experiment. The experiment was performed using the Low Energy Antiproton Ring (LEAR), and was led by Walter Oelert and Mario Macri. Fermilab soon confirmed the CERN findings by producing approximately 100 antihydrogen atoms at their facilities. The antihydrogen atoms created during PS210 and subsequent experiments (at both CERN and Fermilab) were extremely energetic and were not well suited to study. To resolve this hurdle, and to gain a better understanding of antihydrogen, two collaborations were formed in the late 1990s, namely, ATHENA and ATRAP. In 2005, ATHENA disbanded and some of the former members (along with others) formed the ALPHA

Collaboration, which is also based at CERN. The primary goal of these collaborations is the creation of less energetic ("cold") antihydrogen, better suited to study.

In 1999, CERN activated the Antiproton Decelerator, a device capable of decelerating antiprotons from 3500 MeV to 5.3 MeV — still too "hot" to produce study-effective antihydrogen, but a huge leap forward. In late 2002 the ATHENA project announced that they had created the world's first "cold" antihydrogen. The ATRAP project released similar results very shortly thereafter. The antiprotons used in these experiments were cooled by decelerating them with the Antiproton Decelerator, passing them through a thin sheet of foil, and finally capturing them in a Penning–Malmberg trap. The overall cooling process is workable, but highly inefficient; approximately 25 million antiprotons leave the Antiproton Decelerator and roughly 25,000 make it to the Penning–Malmberg trap, which is about 1/1000 or 0.1% of the original amount.

In 2016 a new antiproton decelerator and cooler called ELENA (E Low ENergy Antiproton decelerator) was built. It takes the antiprotons from the antiproton decelerator and cools them to 90 KeV which is "cold" enough to study. More than a hundred antiprotons can be captured per second, a huge improvement, but it would still take several thousand years to make a gram of antimatter.

The antiprotons are still hot when initially trapped. To cool them further, they are mixed into an electron plasma. The electrons in this plasma cool via cyclotron radiation, and then sympathetically cool the antiprotons via Coulomb collisions. Eventually, the electrons are removed by the application of short-duration electric fields, leaving the antiprotons with energies less than 100 meV. While the antiprotons are being cooled in the first trap, a small cloud of positrons is captured from radioactive sodium in a Surko-style positron accumulator. This cloud is then recaptured in a second trap near the antiprotons. Manipulations of the trap electrodes then tip the antiprotons into the positron plasma, where some combine with antiprotons to form antihydrogen. This neutral antihydrogen is unaffected by the electric and magnetic fields used to trap the charged positrons and antiprotons, and within a few microseconds the antihydrogen hits the trap walls, where it annihilates. Some hundreds of millions of antihydrogen atoms have been made in this fashion.

Most of the sought-after high-precision tests of the properties of antihydrogen could only be performed if the antihydrogen were trapped, that is, held in place for a relatively long time. While antihydrogen atoms are electrically neutral, the spins of their component particles produce a magnetic moment. These magnetic moments can interact with an inhomogeneous magnetic field; some of the antihydrogen atoms can be attracted to a magnetic minimum. Such a minimum can be created by a combination of mirror and multipole fields. Antihydrogen can be trapped in such a magnetic minimum (minimum-B) trap; in November 2010, the ALPHA collaboration announced that they had so trapped 38 antihydrogen atoms for about a sixth of a second. This was the first time that neutral antimatter had been trapped.

On 26 April 2011, ALPHA announced that they had trapped 309 antihydrogen atoms, some for as long as 1,000 seconds (about 17 minutes). This was longer than neutral antimatter had ever been trapped before. ALPHA has used these trapped atoms to initiate research into the spectral properties of the antihydrogen.

The biggest limiting factor in the large-scale production of antimatter is the availability of antiprotons. Recent data released by CERN states that, when fully operational, their facilities are capable of producing ten million antiprotons per minute. Assuming a 100% conversion of antiprotons to antihydrogen, it would take 100 billion years to produce 1 gram or 1 mole of antihydrogen (approximately 6.02×10^{23} atoms of anti-hydrogen).

Antihelium

Antihelium-3 nuclei (3He) were first observed in the 1970s in proton–nucleus collision experiments at the Institute for High Energy Physics by Y. Prockoshkin's group (Protvino near Moscow, USSR) and later created in nucleus–nucleus collision experiments. Nucleus–nucleus collisions produce antinuclei through the coalescense of antiprotons and antineutrons created in these reactions. In 2011, the STAR detector reported the observation of artificially created antihelium-4 nuclei (anti-alpha particles) (4He) from such collisions.

Preservation

Antimatter cannot be stored in a container made of ordinary matter because antimatter reacts with any matter it touches, annihilating itself and an equal amount of the container. Antimatter in the form of charged particles can be contained by a combination of electric and magnetic fields, in a device called a Penning trap. This device cannot, however, contain antimatter that consists of uncharged particles, for which atomic traps are used. In particular, such a trap may use the dipole moment (electric or magnetic) of the trapped particles. At high vacuum, the matter or antimatter particles can be trapped and cooled with slightly off-resonant laser radiation using a magneto-optical trap or magnetic trap. Small particles can also be suspended with optical tweezers, using a highly focused laser beam.

In 2011, CERN scientists were able to preserve antihydrogen for approximately 17 minutes.

Cost

Scientists claim that antimatter is the costliest material to make. In 2006, Gerald Smith estimated $250 million could produce 10 milligrams of positrons (equivalent to $25 billion per gram); in 1999, NASA gave a figure of $62.5 trillion per gram of antihydrogen. This is because production is difficult (only very few antiprotons are produced in reactions in particle accelerators), and because there is higher demand for other uses of particle accelerators. According to CERN, it has cost a few hundred million Swiss francs to produce about 1 billionth of a gram (the amount used so far for particle/antiparticle collisions). In comparison, to produce the first atomic weapon, the cost of the Manhattan Project was estimated at $23 billion with inflation during 2007.

Several studies funded by the NASA Institute for Advanced Concepts are exploring whether it might be possible to use magnetic scoops to collect the antimatter that occurs naturally in the Van Allen belt of the Earth, and ultimately, the belts of gas giants, like Jupiter, hopefully at a lower cost per gram.

Uses

Medical

Matter–antimatter reactions have practical applications in medical imaging, such as positron emission tomography (PET). In positive beta decay, a nuclide loses surplus positive charge by emitting a positron (in the same event, a proton becomes a neutron, and a neutrino is also emitted). Nuclides with surplus positive charge are easily made in a cyclotron and are widely generated for medical use. Antiprotons have also been shown within laboratory experiments to have the potential to treat certain cancers, in a similar method currently used for ion (proton) therapy.

Fuel

Isolated and stored anti-matter could be used as a fuel for interplanetary or interstellar travel as part of an antimatter catalyzed nuclear pulse propulsion or other antimatter rocketry, such as the redshift rocket. Since the energy density of antimatter is higher than that of conventional fuels, an antimatter-fueled spacecraft would have a higher thrust-to-weight ratio than a conventional spacecraft.

If matter–antimatter collisions resulted only in photon emission, the entire rest mass of the particles would be converted to kinetic energy. The energy per unit mass (9×10^{16} J/kg) is about 10 orders of magnitude greater than chemical energies, and about 3 orders of magnitude greater than the nuclear potential energy that can be liberated, today, using nuclear fission (about 200 MeV per fission reaction or 8×10^{13} J/kg), and about 2 orders of magnitude greater than the best possible results expected from fusion (about 6.3×10^{14} J/kg for the proton–proton chain). The reaction of 1 kg of antimatter with 1 kg of matter would produce 1.8×10^{17} J (180 petajoules) of energy (by the mass–energy equivalence formula, $E = mc^2$), or the rough equivalent of 43 megatons of TNT – slightly less than the yield of the 27,000 kg Tsar Bomb, the largest thermonuclear weapon ever detonated.

Not all of that energy can be utilized by any realistic propulsion technology because of the nature of the annihilation products. While electron–positron reactions result in gamma ray photons, these are difficult to direct and use for thrust. In reactions between protons and antiprotons, their energy is converted largely into relativistic neutral and charged pions. The neutral pions decay almost immediately (with a half-life of 84 attoseconds) into high-energy photons, but the charged pions decay more slowly (with a half-life of 26 nanoseconds) and can be deflected magnetically to produce thrust.

Note that charged pions ultimately decay into a combination of neutrinos (carrying about 22% of the energy of the charged pions) and unstable charged muons (carrying about 78% of the charged pion energy), with the muons then decaying into a combination of electrons, positrons and neutrinos (cf. muon decay; the neutrinos from this decay carry about 2/3 of the energy of the muons, meaning that from the original charged pions, the total fraction of their energy converted to neutrinos by one route or another would be about $0.22 + (2/3) \cdot 0.78 = 0.74$).

Weapons

Antimatter has been considered as a trigger mechanism for nuclear weapons. A major obstacle is the difficulty of producing antimatter in large enough quantities, and there is no evidence that it will ever be feasible. However, the U.S. Air Force funded studies of the physics of antimatter in the Cold War, and began considering its possible use in weapons, not just as a trigger, but as the explosive itself.

Boson

In quantum mechanics, a boson (/ˈboʊsɒn/, /ˈboʊzɒn/) is a particle that follows Bose–Einstein statistics. Bosons make up one of the two classes of particles, the other being fermions. The name boson was coined by Paul Dirac to commemorate the contribution of the Indian physicist Satyendra Nath Bose in developing, with Einstein, Bose–Einstein statistics—which theorizes the characteristics of elementary particles. Examples of bosons include fundamental particles such as photons, gluons, and W and Z bosons (the four force-carrying gauge bosons of the Standard Model), the recently discovered Higgs boson, and the hypothetical graviton of quantum gravity; composite particles (e.g. mesons and stable nuclei of even mass number such as deuterium (with one proton and one neutron, mass number = 2), helium-4, or lead-208); and some quasiparticles (e.g. Cooper pairs, plasmons, and phonons).

An important characteristic of bosons is that their statistics do not restrict the number of them that occupy the same quantum state. This property is exemplified by helium-4 when it is cooled to become a superfluid. Unlike bosons, two identical fermions cannot occupy the same quantum space. Whereas the elementary particles that make up matter (i.e. leptons and quarks) are fermions, the elementary bosons are force carriers that function as the 'glue' holding matter together. This property holds for all particles with integer spin (s = 0, 1, 2 etc.) as a consequence of the spin–statistics theorem. When a gas of Bose particles is cooled down to temperatures very close to absolute zero then the kinetic energy of the particles decreases to a negligible amount and they condense into a lowest energy level state. This state is called Bose-Einstein condensation. It is believed that this property is the explanation of superfluidity.

Satyendra Nath Bose

Types

Bosons may be either elementary, like photons, or composite, like mesons.

While most bosons are composite particles, in the Standard Model there are five bosons which are elementary:

- the four gauge bosons ($\gamma \cdot g \cdot Z \cdot W\pm$)
- the only scalar boson (the Higgs boson (Ho))

Additionally, the graviton (G) is a hypothetical elementary particle not incorporated in the Standard Model. If it exists, a graviton must be a boson, and could conceivably be a gauge boson.

Composite bosons are important in superfluidity and other applications of Bose–Einstein condensates. When a gas of Bose particles is cooled to temperatures very close to absolute zero its kinetic energy decreases up to a negligible amount then the particles would condense into the lowest energy state. This phenomenon is known as Bose-Einstein condensation and it is believed that this phenomenon is the secret behind superfluidity of liquids.

Properties

Bosons differ from fermions, which obey Fermi–Dirac statistics. Two or more identical fermions cannot occupy the same quantum state.

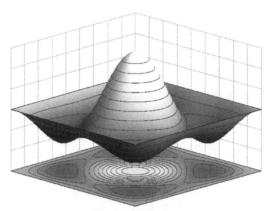

Symmetric wavefunction for a (bosonic) 2-particle state in an infinite square well potential.

Since bosons with the same energy can occupy the same place in space, bosons are often force carrier particles. Fermions are usually associated with matter (although in quantum mechanics the distinction between the two concepts is not clear cut)

Bosons are particles which obey Bose–Einstein statistics: when one swaps two bosons (of the same species), the wavefunction of the system is unchanged. Fermions, on the other hand, obey Fermi–Dirac statistics and the Pauli exclusion principle: two fermions cannot occupy the same quantum state, resulting in a "rigidity" or "stiffness" of matter which includes fermions. Thus fermions are sometimes said to be the constituents of matter, while bosons are said to be the particles that transmit interactions (force carriers), or the constituents of radiation. The quantum fields of bosons are bosonic fields, obeying canonical commutation relations.

The properties of lasers and masers, superfluid helium-4 and Bose–Einstein condensates are all consequences of statistics of bosons. Another result is that the spectrum of a photon gas in thermal equilibrium is a Planck spectrum, one example of which is black-body radiation; another is the thermal radiation of the opaque early Universe seen today as microwave background radiation. Interactions between elementary particles are called fundamental interactions. The fundamental interactions of virtual bosons with real particles result in all forces we know.

All known elementary and composite particles are bosons or fermions, depending on their spin: particles with half-integer spin are fermions; particles with integer spin are bosons. In the framework of nonrelativistic quantum mechanics, this is a purely empirical observation. However, in relativistic quantum field theory, the spin–statistics theorem shows that half-integer spin particles cannot be bosons and integer spin particles cannot be fermions.

In large systems, the difference between bosonic and fermionic statistics is only apparent at large densities—when their wave functions overlap. At low densities, both types of statistics are well approximated by Maxwell–Boltzmann statistics, which is described by classical mechanics.

Elementary Bosons

All observed elementary particles are either fermions or bosons. The observed elementary bosons are all gauge bosons: photons, W and Z bosons, gluons, except the Higgs boson which is a scalar boson.

- Photons are the force carriers of the electromagnetic field.

- W and Z bosons are the force carriers which mediate the weak force.

- Gluons are the fundamental force carriers underlying the strong force.

- Higgs bosons give W and Z bosons mass via the Higgs mechanism. Their existence was confirmed by CERN on 14 March 2013.

Finally, many approaches to quantum gravity postulate a force carrier for gravity, the graviton, which is a boson of spin plus or minus two.

Composite Bosons

Composite particles (such as hadrons, nuclei, and atoms) can be bosons or fermions depending on their constituents. More precisely, because of the relation between spin and statistics, a particle containing an even number of fermions is a boson, since it has integer spin.

Examples include the following:

- Any meson, since mesons contain one quark and one antiquark.

- The nucleus of a carbon-12 atom, which contains 6 protons and 6 neutrons.

- The helium-4 atom, consisting of 2 protons, 2 neutrons and 2 electrons.

The number of bosons within a composite particle made up of simple particles bound with a potential has no effect on whether it is a boson or a fermion.

To Which States Can Bosons Crowd?

Bose–Einstein statistics encourages identical bosons to crowd into one quantum state, but not any state is necessarily convenient for it. Aside of statistics, bosons can interact – for example, helium-4 atoms are repulsed by intermolecular force on a very close approach, and if one hypothesizes their condensation in a spatially-localized state, then gains from the statistics cannot overcome a prohibitive force potential. A spatially-delocalized state (i.e. with low $|\psi(x)|$) is preferable: if the number density of the condensate is about the same as in ordinary liquid or solid state, then the repulsive potential for the N-particle condensate in such state can be no higher than for a liquid or a crystalline lattice of the same N particles described without quantum statistics. Thus, Bose–Einstein statistics for a material particle is not a mechanism to bypass physical restrictions on the density of the corresponding substance, and superfluid liquid helium has the density comparable to the density of ordinary liquid matter. Spatially-delocalized states also permit for a low momentum according to uncertainty principle, hence for low kinetic energy; this is why superfluidity and superconductivity are usually observed in low temperatures.

Photons do not interact with themselves and hence do not experience this difference in states where to crowd.

References

- Friedel Weinert (2004). The Scientist as Philosopher: Philosophical Consequences of Great Scientific Discoveries. Springer. p. 43. ISBN 978-3-540-20580-7.

- Sylvie Braibant; Giorgio Giacomelli; Maurizio Spurio (2012). Particles and Fundamental Interactions: An Introduction to Particle Physics (2nd ed.). Springer. p. 384. ISBN 978-94-007-2463-1.

- J.D. Barrow (1997) [1994]. "The Singularity and Other Problems". The Origin of the Universe (Reprint ed.). Basic Books. ISBN 978-0-465-05314-8.

- R.C. Olby; G.N. Cantor (1996). Companion to the History of Modern Science. Taylor & Francis. p. 673. ISBN 0-415-14578-3.

- M. Gell-Mann (1995). The Quark and the Jaguar: Adventures in the Simple and the Complex. Henry Holt and Co. p. 180. ISBN 978-0-8050-7253-2.

- M. Riordan (1987). The Hunting of the Quark: A True Story of Modern Physics. Simon & Schuster. p. 210. ISBN 978-0-671-50466-3.

- Part III of M.E. Peskin, D.V. Schroeder (1995). An Introduction to Quantum Field Theory. Addison–Wesley. ISBN 0-201-50397-2.

- R.P. Feynman (1985). QED: The Strange Theory of Light and Matter (1st ed.). Princeton University Press. pp. 136–137. ISBN 0-691-08388-6.

- Role as gauge boson and polarization section 5.1 inAitchison, I.J.R.; Hey, A.J.G. (1993). Gauge Theories in Particle Physics. IOP Publishing. ISBN 0-85274-328-9.

- Halliday, David; Resnick, Robert; Walker, Jerl (2005), Fundamental of Physics (7th ed.), USA: John Wiley and Sons, Inc., ISBN 0-471-23231-9

- See section 1.6 in Alonso, M.; Finn, E.J. (1968). Fundamental University Physics Volume III: Quantum and Statistical Physics. Addison-Wesley. ISBN 0-201-00262-0.

6

Technologies Related to Particle Physics

Particle physics has a number of technologies associated with it; some of these are particle accelerator, accelerator physics and particle physics experiments. Particle accelerators are machines that use electromagnetic fields to boost charged particles in order to include them in well-defined beams. This chapter elucidates the main technologies related to particle physics.

Particle Accelerator

A particle accelerator is a machine that uses electromagnetic fields to propel charged particles to nearly light speed and to contain them in well-defined beams. Large accelerators are used in particle physics as colliders (e.g. the LHC at CERN, KEKB at KEK in Japan, RHIC at Brookhaven National Laboratory, and Tevatron at Fermilab), or as synchrotron light sources for the study of condensed matter physics. Smaller particle accelerators are used in a wide variety of applications, including particle therapy for oncological purposes, radioisotope production for medical diagnostics, ion implanters for manufacture of semiconductors, and accelerator mass spectrometers for measurements of rare isotopes such as radiocarbon. There are currently more than 30,000 accelerators in operation around the world.

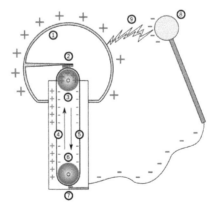

Sketch of an electrostatic Van de Graaff accelerator

There are two basic classes of accelerators: electrostatic and electrodynamic (or electromagnetic) accelerators. *Electrostatic* accelerators use static electric fields to accelerate particles. A small-scale example of this class is the cathode ray tube in an ordinary old television set. Other examples are the Cockcroft–Walton generator and the Van de Graaff generator. The achievable kinetic energy for particles in these devices is limited by electrical breakdown. *Electrodynamic* or *electromagnetic* accelerators, on the other hand, use changing electromagnetic fields (either magnetic induction or oscillating radio frequency fields) to accelerate particles and reduce the breakdown problem. This class, which was first developed in the 1920s, is the basis for most modern accelerator concepts and large-scale facilities.

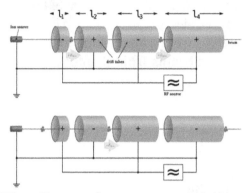

Sketch of the Ising/Wideröe linear accelerator concept, employing oscillating fields (1928)

Rolf Wideröe, Gustav Ising, Leó Szilárd, Max Steenbeck, and Ernest Lawrence are considered pioneers of this field, conceiving and building the first operational linear particle accelerator, the betatron, and the cyclotron.

Because colliders can give evidence of the structure of the subatomic world, accelerators were commonly referred to as atom smashers in the 20th century. Despite the fact that most accelerators (but not ion facilities) actually propel subatomic particles, the term persists in popular usage when referring to particle accelerators in general.

Uses

Beamlines leading from the Van de Graaff accelerator to various experiments, in the basement of the Jussieu Campus in Paris.

Breakdown of the cumulative number of industrial particle accelerators according to their applications.

The now disused Koffler particle accelerator at the Weizmann Institute, Rehovot, Israel.

Beams of high-energy particles are useful for both fundamental and applied research in the sciences, and also in many technical and industrial fields unrelated to fundamental research. It has been estimated that there are approximately 30,000 accelerators worldwide. Of these, only about 1% are research machines with energies above 1 GeV, while about 44% are for radiotherapy, 41% for ion implantation, 9% for industrial processing and research, and 4% for biomedical and other low-energy research. The bar graph shows the breakdown of the number of industrial accelerators according to their applications. The numbers are based on 2012 statistics available from various sources, including production and sales data published in presentations or market surveys, and data provided by a number of manufacturers.

High-energy Physics

For the most basic inquiries into the dynamics and structure of matter, space, and time, physicists seek the simplest kinds of interactions at the highest possible energies. These typically entail particle energies of many GeV, and the interactions of the simplest kinds of particles: leptons (e.g. electrons and positrons) and quarks for the matter, or photons and gluons for the field quanta. Since isolated quarks are experimentally unavailable due to color confinement, the simplest available experiments involve the interactions of, first, leptons with each other, and second, of leptons with nucleons, which are composed of quarks and gluons. To study the collisions of quarks with each other, scientists resort to collisions of nucleons, which at high energy may be usefully considered as essentially 2-body interactions of the quarks and gluons of which they are composed. Thus elementary particle physicists tend to use machines creating beams of electrons, positrons, protons, and antiprotons, interacting with each other or with the simplest nuclei (e.g., hydrogen or deuterium) at the highest possible energies, generally hundreds of GeV or more.

The largest and highest energy particle accelerator used for elementary particle physics is the Large Hadron Collider (LHC) at CERN (which came on-line in mid-November 2009).

Nuclear Physics and Isotope Production

Nuclear physicists and cosmologists may use beams of bare atomic nuclei, stripped of electrons, to investigate the structure, interactions, and properties of the nuclei themselves, and of condensed matter at extremely high temperatures and densities, such as might have occurred in the first moments of the Big Bang. These investigations often involve collisions of heavy nuclei – of atoms like iron or gold – at energies of several GeV per nucleon. The largest such particle accelerator is the Relativistic Heavy Ion Collider (RHIC) at Brookhaven National Laboratory.

Particle accelerators can also produce proton beams, which can produce proton-rich medical or research isotopes as opposed to the neutron-rich ones made in fission reactors; however, recent work has shown how to make ^{99}Mo, usually made in reactors, by accelerating isotopes of hydrogen, although this method still requires a reactor to produce tritium. An example of this type of machine is LANSCE at Los Alamos.

Synchrotron Radiation

Besides being of fundamental interest, high energy electrons may be coaxed into emitting extremely bright and coherent beams of high energy photons via synchrotron radiation, which have numerous uses in the study of atomic structure, chemistry, condensed matter physics, biology, and technology. A large number of synchrotron light sources exist worldwide. Examples in the US are SSRL and LCLS at SLAC National Accelerator Laboratory, APS at Argonne National Laboratory, ALS at Lawrence Berkeley National Laboratory, and NSLS at Brookhaven National Laboratory. The ESRF in Grenoble, France has been used to extract detailed 3-dimensional images of insects trapped in amber. Thus there is a great demand for electron accelerators of moderate (GeV) energy and high intensity.

Low-energy Machines and Particle Therapy

Everyday examples of particle accelerators are cathode ray tubes found in television sets and X-ray generators. These low-energy accelerators use a single pair of electrodes with a DC voltage of a few thousand volts between them. In an X-ray generator, the target itself is one of the electrodes. A low-energy particle accelerator called an ion implanter is used in the manufacture of integrated circuits.

At lower energies, beams of accelerated nuclei are also used in medicine as particle therapy, for the treatment of cancer.

DC accelerator types capable of accelerating particles to speeds sufficient to cause nuclear reactions are Cockcroft-Walton generators or voltage multipliers, which convert AC to high voltage DC, or Van de Graaff generators that use static electricity carried by belts.

Electrostatic Particle Accelerators

A Cockcroft-Walton generator (Philips, 1937), residing in Science Museum (London).

A 1960s single stage 2 MeV linear Van de Graaff accelerator, here opened for maintenance

Historically, the first accelerators used simple technology of a single static high voltage to accelerate charged particles. The charged particle was accelerated through an evacuated tube with an electrode at either end, with the static potential across it. Since the particle passed only once through the potential difference, the output energy was limited to the accelerating voltage of the machine. While this method is still extremely popular today, with the electrostatic accelerators greatly out-numbering any other type, they are more suited to lower energy studies owing to the practical voltage limit of about 1 MV for air insulated machines, or 30 MV when the accelerator is operated in a tank of pressurized gas with high dielectric strength, such as sulfur hexafluoride. In a *tandem accelerator* the potential is used twice to accelerate the particles, by reversing the charge of the particles while they are inside the terminal. This is possible with the acceleration of atomic nuclei by using anions (negatively charged ions), and then passing the beam through a thin foil to strip electrons off the anions inside the high voltage terminal, converting them to cations (positively charged ions), which are accelerated again as they leave the terminal.

The two main types of electrostatic accelerator are the Cockcroft-Walton accelerator, which uses a diode-capacitor voltage multiplier to produce high voltage, and the Van de Graaff accelerator, which uses a moving fabric belt to carry charge to the high voltage electrode. Although electrostatic accelerators accelerate particles along a straight line, the term linear accelerator is more often used for accelerators that employ oscillating rather than static electric fields.

Electrodynamic (Electromagnetic) Particle Accelerators

Due to the high voltage ceiling imposed by electrical discharge, in order to accelerate particles to higher energies, techniques involving dynamic fields rather than static fields are used. Electrodynamic acceleration can arise from either of two mechanisms: non-resonant magnetic induction, or resonant circuits or cavities excited by oscillating RF fields. Electrodynamic accelerators can be *linear*, with particles accelerating in a straight line, or *circular*, using magnetic fields to bend particles in a roughly circular orbit.

Magnetic Induction Accelerators

Magnetic induction accelerators accelerate particles by induction from an increasing magnetic field, as if the particles were the secondary winding in a transformer. The increasing magnetic field creates a circulating electric field which can be configured to accelerate the particles. Induction accelerators can be either linear or circular.

Linear Induction Accelerators

Linear induction accelerators utilize ferrite-loaded, non-resonant induction cavities. Each cavity can be thought of as two large washer-shaped disks connected by an outer cylindrical tube. Between the disks is a ferrite toroid. A voltage pulse applied between the two disks causes an increasing magnetic field which inductively couples power into the charged particle beam.

The linear induction accelerator was invented by Christofilos in the 1960s. Linear induction accelerators are capable of accelerating very high beam currents (>1000 A) in a single short pulse. They have been used to generate X-rays for flash radiography (e.g. DARHT at LANL), and have been considered as particle injectors for magnetic confinement fusion and as drivers for free electron lasers.

Betatrons

The Betatron is circular magnetic induction accelerator, invented by Donald Kerst in 1940 for accelerating electrons. The concept originates ultimately from Norwegian-Ger-man scientist Rolf Wideröe. These machines, like synchrotrons, use a donut-shaped ring magnet with a cyclically increasing B field, but accelerate the particles by induction from the increasing magnetic field, as if they were the secondary winding in a transformer, due to the changing magnetic flux through the orbit.

Achieving constant orbital radius while supplying the proper accelerating electric field requires that the magnetic flux linking the orbit be somewhat independent of the magnetic field on the orbit, bending the particles into a constant radius curve. These machines have in practice been limited by the large radiative losses suffered by the electrons moving at nearly the speed of light in a relatively small radius orbit.

RF Linear Accelerators

Modern superconducting radio frequency, multicell linear accelerator component.

In a linear particle accelerator (linac), particles are accelerated in a straight line with a target of interest at one end. They are often used to provide an initial low-energy kick to particles before they are injected into circular accelerators. The longest linac in the world is the Stanford Linear Accelerator, SLAC, which is 3 km (1.9 mi) long. SLAC is an electron-positron collider.

Linear high-energy accelerators use a linear array of plates (or drift tubes) to which an alternating high-energy field is applied. As the particles approach a plate they are accelerated towards it by an opposite polarity charge applied to the plate. As they pass through a hole in the plate, the polarity is switched so that the plate now repels them and they are now accelerated by it towards the next plate. Normally a stream of "bunches" of particles are accelerated, so a carefully controlled AC voltage is applied to each plate to continuously repeat this process for each bunch.

As the particles approach the speed of light the switching rate of the electric fields becomes so high that they operate at radio frequencies, and so microwave cavities are used in higher energy machines instead of simple plates.

Linear accelerators are also widely used in medicine, for radiotherapy and radiosurgery. Medical grade linacs accelerate electrons using a klystron and a complex bending magnet arrangement which produces a beam of 6-30 MeV energy. The electrons can be used directly or they can be collided with a target to produce a beam of X-rays. The reliability, flexibility and accuracy of the radiation beam produced has largely supplanted the older use of cobalt-60 therapy as a treatment tool.

Circular or Cyclic RF Accelerators

In the circular accelerator, particles move in a circle until they reach sufficient energy. The particle track is typically bent into a circle using electromagnets. The advantage of circular accelerators over linear accelerators (*linacs*) is that the ring topology allows continuous acceleration, as the particle can transit indefinitely. Another advantage is that a circular accelerator is smaller than a linear accelerator of comparable power (i.e. a linac would have to be extremely long to have the equivalent power of a circular accelerator).

Depending on the energy and the particle being accelerated, circular accelerators suffer a disadvantage in that the particles emit synchrotron radiation. When any charged particle is accelerated, it emits electromagnetic radiation and secondary emissions. As a particle traveling in a circle is always accelerating towards the center of the circle, it continuously radiates towards the tangent of the circle. This radiation is called synchrotron light and depends highly on the mass of the accelerating particle. For this reason, many high energy electron accelerators are linacs. Certain accelerators (synchrotrons) are however built specially for producing synchrotron light (X-rays).

Since the special theory of relativity requires that matter always travels slower than the speed of light in a vacuum, in high-energy accelerators, as the energy increases the particle speed approaches the speed of light as a limit, but never attains it. Therefore, particle physicists do not generally think in terms of speed, but rather in terms of a particle's energy or momentum, usually measured in electron volts (eV). An important principle for circular accelerators, and particle beams in general, is that the curvature of the particle trajectory is proportional to the particle charge and to the magnetic field, but inversely proportional to the (typically relativistic) momentum.

Cyclotrons

The earliest operational circular accelerators were cyclotrons, invented in 1929 by Ernest O. Lawrence at the University of California, Berkeley. Cyclotrons have a single pair of hollow 'D'-shaped plates to accelerate the particles and a single large dipole magnet to bend their path into a circular orbit. It is a characteristic property of charged particles in a uniform and constant magnetic field B that they orbit with a constant period, at a frequency called the cyclotron frequency, so long as their speed is small compared to the speed of light c. This means that the accelerating D's of a cyclotron can be driven at a constant frequency by a radio frequency (RF) accelerating power source, as the beam spirals outwards continuously. The particles are injected in the centre of the magnet and are extracted at the outer edge at their maximum energy.

Lawrence's 60 inch cyclotron, with magnet poles 60 inches (5 feet, 1.5 meters) in diameter, at the University of California Lawrence Radiation Laboratory, Berkeley, in August, 1939, the most powerful accelerator in the world at the time. Glenn T. Seaborg and Edwin M. McMillan *(right)* used it to discover plutonium, neptunium and many other transuranic elements and isotopes, for which they received the 1951 Nobel Prize in chemistry.

Cyclotrons reach an energy limit because of relativistic effects whereby the particles effectively become more massive, so that their cyclotron frequency drops out of synch with the accelerating RF. Therefore, simple cyclotrons can accelerate protons only to an energy of around 15 million electron volts (15 MeV, corresponding to a speed of roughly 10% of c), because the protons get out of phase with the driving electric field. If accelerated further, the beam would continue to spiral outward to a larger radius but the particles would no longer gain enough speed to complete the larger circle in step with the accelerating RF. To accommodate relativistic effects the magnetic field needs to be increased to higher radii like it is done in isochronous cyclotrons. An example of an isochronous cyclotron is the PSI Ring cyclotron in Switzerland, which provides protons at the energy of 590 MeV which corresponds to roughly 80% of the speed of light. The advantage of such a cyclotron is the maximum achievable extracted proton current which is currently 2.2 mA. The energy and current correspond to 1.3 MW beam power which is the highest of any accelerator currently existing.

Synchrocyclotrons and Isochronous Cyclotrons

A classic cyclotron can be modified to increase its energy limit. The historically first approach was the synchrocyclotron, which accelerates the particles in bunches. It uses a constant magnetic field B, but reduces the accelerating field's frequency so as to keep the particles in step as they spiral outward, matching their mass-dependent cyclotron resonance frequency. This approach suffers from low average beam intensity due to the bunching, and again from the need for a huge magnet of large radius and constant field over the larger orbit demanded by high energy.

A magnet in the synchrocyclotron at the Orsay proton therapy center

The second approach to the problem of accelerating relativistic particles is the isochronous cyclotron. In such a structure, the accelerating field's frequency (and the cyclotron resonance frequency) is kept constant for all energies by shaping the magnet poles so to increase magnetic field with radius. Thus, all particles get accelerated in isochronous time intervals. Higher energy particles travel a shorter distance in each orbit than they would in a classical cyclotron, thus remaining in phase with the accelerating field. The advantage of the isochronous cyclotron is that it can deliver continuous beams of higher average intensity, which is useful for some applications. The main disadvantages are the size and cost of the large magnet needed, and the difficulty in achieving the high magnetic field values required at the outer edge of the structure.

Synchrocyclotrons have not been built since the isochronous cyclotron was developed.

Synchrotrons

To reach still higher energies, with relativistic mass approaching or exceeding the rest mass of the particles (for protons, billions of electron volts or GeV), it is necessary to use a synchrotron. This is an accelerator in which the particles are accelerated in a ring of constant radius. An immediate advantage over cyclotrons is that the magnetic field need only be present over the actual region of the particle orbits, which is much narrower than that of the ring. (The largest cyclotron built in the US had a 184-inch-diameter (4.7 m) magnet pole, whereas the diameter of synchrotrons such as the LEP and LHC is nearly 10 km. The aperture of the two beams of the LHC is of the order of a millimeter.)

However, since the particle momentum increases during acceleration, it is necessary to turn up the magnetic field B in proportion to maintain constant curvature of the orbit. In consequence, synchrotrons cannot accelerate particles continuously, as cyclotrons can, but must operate cyclically, supplying particles in bunches, which are delivered to a target or an external beam in beam "spills" typically every few seconds.

Aerial photo of the Tevatron at Fermilab, which resembles a figure eight. The main accelerator is the ring above; the one below (about half the diameter, despite appearances) is for preliminary acceleration, beam cooling and storage, etc.

Since high energy synchrotrons do most of their work on particles that are already traveling at nearly the speed of light c, the time to complete one orbit of the ring is nearly constant, as is the frequency of the RF cavity resonators used to drive the acceleration.

In modern synchrotrons, the beam aperture is small and the magnetic field does not cover the entire area of the particle orbit as it does for a cyclotron, so several necessary functions can be separated. Instead of one huge magnet, one has a line of hundreds of bending magnets, enclosing (or enclosed by) vacuum connecting pipes. The design of synchrotrons was revolutionized in the early 1950s with the discovery of the strong focusing concept. The focusing of the beam is handled independently by specialized quadrupole magnets, while the acceleration itself is accomplished in separate RF sections, rather similar to short linear accelerators. Also, there is no necessity that cyclic machines be circular, but rather the beam pipe may have straight sections between magnets where beams may collide, be cooled, etc. This has developed into an entire separate subject, called "beam physics" or "beam optics".

More complex modern synchrotrons such as the Tevatron, LEP, and LHC may deliver the particle bunches into storage rings of magnets with constant B, where they can continue to orbit for long periods for experimentation or further acceleration. The highest-energy machines such as the Tevatron and LHC are actually accelerator complexes, with a cascade of specialized elements in series, including linear accelerators for initial beam creation, one or more low energy synchrotrons to reach intermediate energy, storage rings where beams can be accumulated or "cooled" (reducing the magnet aperture required and permitting tighter focusing; see beam cooling), and a last large ring for final acceleration and experimentation.

Segment of an electron synchrotron at DESY

Electron Synchrotrons

Circular electron accelerators fell somewhat out of favor for particle physics around the time that SLAC's linear particle accelerator was constructed, because their synchrotron losses were considered economically prohibitive and because their beam intensity was lower than for the unpulsed linear machines. The Cornell Electron Synchrotron, built at low cost in the late 1970s, was the first in a series of high-energy circular electron accelerators built for fundamental particle physics, the last being LEP, built at CERN, which was used from 1989 until 2000.

A large number of electron synchrotrons have been built in the past two decades, as part of synchrotron light sources that emit ultraviolet light and X rays.

Storage Rings

For some applications, it is useful to store beams of high energy particles for some time (with modern high vacuum technology, up to many hours) without further acceleration. This is especially true for colliding beam accelerators, in which two beams moving in opposite directions are made to collide with each other, with a large gain in effective collision energy. Because relatively few collisions occur at each pass through the intersection point of the two beams, it is customary to first accelerate the beams to the desired energy, and then store them in storage rings, which are essentially synchrotron rings of magnets, with no significant RF power for acceleration.

Synchrotron Radiation Sources

Some circular accelerators have been built to deliberately generate radiation (called synchrotron light) as X-rays also called synchrotron radiation, for example the Diamond Light Source which has been built at the Rutherford Appleton Laboratory in England or the Advanced Photon Source at Argonne National Laboratory in Illinois, USA. High-energy X-rays are useful for X-ray spectroscopy of proteins or X-ray absorption fine structure (XAFS), for example.

Synchrotron radiation is more powerfully emitted by lighter particles, so these accelerators are invariably electron accelerators. Synchrotron radiation allows for better imaging as researched and developed at SLAC's SPEAR.

FFAG Accelerators

Fixed-Field Alternating Gradient accelerators (FFAG)s, in which a very strong radial field gradient, combined with strong focusing, allows the beam to be confined to a narrow ring, are an extension of the isochronous cyclotron idea that is lately under development. They use RF accelerating sections between the magnets, and so are isochronous for relativistic particles like electrons (which achieve essentially the speed of light at only a few MeV), but only over a limited energy range for protons and heavier particles at sub-relativistic energies. Like the isochronous cyclotrons, they achieve continuous beam operation, but without the need for a huge dipole bending magnet covering the entire radius of the orbits.

History

Ernest Lawrence's first cyclotron was a mere 4 inches (100 mm) in diameter. Later, in 1939, he built a machine with a 60-inch diameter pole face, and planned one with a 184-inch diameter in 1942, which was, however, taken over for World War II-related work connected with uranium isotope separation; after the war it continued in service for research and medicine over many years.

The first large proton synchrotron was the Cosmotron at Brookhaven National Laboratory, which accelerated protons to about 3 GeV (1953–1968). The Bevatron at Berkeley, completed in 1954, was specifically designed to accelerate protons to sufficient energy to create antiprotons, and verify the particle-antiparticle symmetry of nature, then only theorized. The Alternating Gradient Synchrotron (AGS) at Brookhaven (1960–) was the first large synchrotron with alternating gradient, "strong focusing" magnets, which greatly reduced the required aperture of the beam, and correspondingly the size and cost of the bending magnets. The Proton Synchrotron, built at CERN (1959–), was the first major European particle accelerator and generally similar to the AGS.

The Stanford Linear Accelerator, SLAC, became operational in 1966, accelerating electrons to 30 GeV in a 3 km long waveguide, buried in a tunnel and powered by hundreds of large klystrons. It is still the largest linear accelerator in existence, and has been upgraded with the addition of storage rings and an electron-positron collider facility. It is also an X-ray and UV synchrotron photon source.

The Fermilab Tevatron has a ring with a beam path of 4 miles (6.4 km). It has received several upgrades, and has functioned as a proton-antiproton collider until it was shut down due to budget cuts on September 30, 2011. The largest circular accelerator ever built was the LEP synchrotron at CERN with a circumference 26.6 kilometers, which was an electron/positron collider. It achieved an energy of 209 GeV before it was dismantled in 2000 so that the underground tunnel could be used for the Large Hadron Collider (LHC). The LHC is a proton collider, and currently the world's largest and highest-energy accelerator, achieving 7 TeV energy per beam (14 TeV in total).

The aborted Superconducting Super Collider (SSC) in Texas would have had a circumference of 87 km. Construction was started in 1991, but abandoned in 1993. Very large circular accelerators are invariably built in underground tunnels a few metres wide to minimize the disruption and cost of building such a structure on the surface, and to provide shielding against intense secondary radiations that occur, which are extremely penetrating at high energies.

Current accelerators such as the Spallation Neutron Source, incorporate superconducting cryomodules. The Relativistic Heavy Ion Collider, and Large Hadron Collider also make use of superconducting magnets and RF cavity resonators to accelerate particles.

Targets and Detectors

The output of a particle accelerator can generally be directed towards multiple lines of experiments, one at a given time, by means of a deviating electromagnet. This makes it possible to operate multiple experiments without needing to move things around or shutting down the entire accelerator beam. Except for synchrotron radiation sources, the purpose of an accelerator is to generate high-energy particles for interaction with matter.

This is usually a fixed target, such as the phosphor coating on the back of the screen in the case of a television tube; a piece of uranium in an accelerator designed as a neutron source; or a tungsten target for an X-ray generator. In a linac, the target is simply fitted to the end of the accelerator. The particle track in a cyclotron is a spiral outwards from the centre of the circular machine, so the accelerated particles emerge from a fixed point as for a linear accelerator.

For synchrotrons, the situation is more complex. Particles are accelerated to the desired energy. Then, a fast acting dipole magnet is used to switch the particles out of the circular synchrotron tube and towards the target.

A variation commonly used for particle physics research is a collider, also called a *storage ring collider*. Two circular synchrotrons are built in close proximity – usually on top of each other and using the same magnets (which are then of more complicated design to accommodate both beam tubes). Bunches of particles travel in opposite directions around the two accelerators and collide at intersections between them. This can increase the energy enormously; whereas in a fixed-target experiment the energy available to produce new particles is proportional to the square root of the beam energy, in a collider the available energy is linear.

Higher Energies

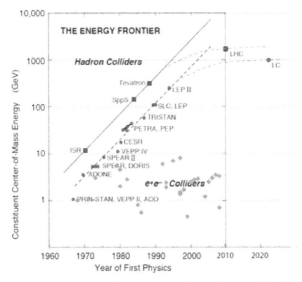

A Livingston chart depicting progress in collision energy through 2010. The LHC is the largest collision energy to date, but also represents the first break in the log-linear trend.

At present the highest energy accelerators are all circular colliders, but both hadron accelerators and electron accelerators are running into limits. Higher energy hadron and ion cyclic accelerators will require accelerator tunnels of larger physical size due to the increased beam rigidity.

For cyclic electron accelerators, a limit on practical bend radius is placed by synchrotron radiation losses and the next generation will probably be linear accelerators 10 times the current length. An example of such a next generation electron accelerator is the 40 km long International Linear Collider, due to be constructed between 2015–2020.

It is believed that plasma wakefield acceleration in the form of electron-beam 'afterburners' and standalone laser pulsers might be able to provide dramatic increases in efficiency over RF accelerators within two to three decades. In plasma wakefield accelerators, the beam cavity is filled with a plasma (rather than vacuum). A short pulse of electrons or laser light either constitutes or immediately precedes the particles that are being accelerated. The pulse disrupts the plasma, causing the charged particles in the plasma to integrate into and move toward the rear of the bunch of particles that are being accelerated. This process transfers energy to the particle bunch, accelerating it further, and continues as long as the pulse is coherent.

Energy gradients as steep as 200 GeV/m have been achieved over millimeter-scale distances using laser pulsers and gradients approaching 1 GeV/m are being produced on the multi-centimeter-scale with electron-beam systems, in contrast to a limit of about 0.1 GeV/m for radio-frequency acceleration alone. Existing electron accelerators such as SLAC could use electron-beam afterburners to greatly increase the energy of their particle beams, at the cost of beam intensity. Electron systems in general can provide tightly collimated, reliable beams; laser systems may offer more power and compactness. Thus, plasma wakefield accelerators could be used – if technical issues can be resolved – to both increase the maximum energy of the largest accelerators and to bring high energies into university laboratories and medical centres.

Higher than 0.25 GeV/m gradients have been achieved by a dielectric laser accelerator, which may present another viable approach to building compact high-energy accelerators.

Black Hole Production and Public Safety Concerns

In the future, the possibility of black hole production at the highest energy accelerators may arise if certain predictions of superstring theory are accurate. This and other exotic possibilities have led to public safety concerns that have been widely reported in connection with the LHC, which began operation in 2008. The various possible dangerous scenarios have been assessed as presenting "no conceivable danger" in the latest risk assessment produced by the LHC Safety Assessment Group. If black holes are produced, it is theoretically predicted that such small black holes should evaporate extremely quickly via Bekenstein-Hawking radiation, but which is as yet experimentally unconfirmed. If colliders can produce black holes, cosmic rays (and particularly ultra-high-energy cosmic rays, UHECRs) must have been producing them for eons, but they have yet to harm anybody. It has been argued that to conserve energy and momentum, any black holes created in a collision between an UHECR and local matter would necessarily be produced moving at relativistic speed with respect to the Earth, and should escape into space, as their accretion and growth rate should be very slow, while black holes produced in colliders (with components of equal mass) would have some chance of having a velocity less than Earth escape velocity, 11.2 km per sec, and would be liable to capture and subsequent growth. Yet even on such scenarios the collisions of UHECRs with white dwarfs and neutron stars would lead to their rapid destruction, but these bodies are observed to be common astronomical objects. Thus if stable micro black holes should be produced, they must grow far too slowly to cause any noticeable macroscopic effects within the natural lifetime of the solar system.

Accelerator Operator

An accelerator operator controls the operation of a particle accelerator used in research experiments, reviews an experiment schedule to determine experiment parameters specified by an experimenter (physicist), adjust particle beam parameters such as aspect ratio, current intensity, and position on target, communicates with and assists accelerator maintenance personnel to ensure readiness of support systems, such as vacuum, magnet power supplies and controls, low conductivity water or LCW cooling, and radiofrequency power supplies and controls, and maintains a record of accelerator related events.

Accelerator Physics

Accelerator physics is a branch of applied physics, concerned with designing, building and operating particle accelerators. As such, it can be described as the study of motion, manipulation and observation of relativistic charged particle beams and their interaction with accelerator structures by electromagnetic fields.

It is also related to other fields:

- Microwave engineering (for acceleration/deflection structures in the radio frequency range).

- Optics with an emphasis on geometrical optics (beam focusing and bending) and laser physics (laser-particle interaction).

- Computer technology with an emphasis on digital signal processing; e.g., for automated manipulation of the particle beam.

The experiments conducted with particle accelerators are not regarded as part of accelerator physics, but belong (according to the objectives of the experiments) to, e.g., particle physics, nuclear physics, condensed matter physics or materials physics. The types of experiments done at a particular accelerator facility are determined by characteristics of the generated particle beam such as average energy, particle type, intensity, and dimensions.

Acceleration and Interaction of Particles with RF Structures

Superconducting niobium cavity for acceleration of ultrarelativistic particles from the TESLA project

While it is possible to accelerate charged particles using electrostatic fields, like in a Cockcroft-Walton voltage multiplier, this method has limits given by electrical breakdown at high voltages. Furthermore, due to electrostatic fields being conservative, the maximum voltage limits the kinetic energy that is applicable to the particles.

To circumvent this problem, linear particle accelerators operate using time-varying fields. To control this fields using hollow macroscopic structures through which the particles are passing (wavelength restrictions), the frequency of such acceleration fields is located in the radio frequency region of the electromagnetic spectrum.

The space around a particle beam is evacuated to prevent scattering with gas atoms, requiring it to be enclosed in a vacuum chamber (or *beam pipe*). Due to the strong electromagnetic fields that follow the beam, it is possible for it to interact with any electrical impedance in the walls of the beam pipe. This may be in the form of a resistive impedance (i.e., the finite resistivity of the beam pipe material) or an inductive/capacitive impedance (due to the geometric changes in the beam pipe's cross section).

These impedances will induce *wakefields* (a strong warping of the electromagnetic field of the beam) that can interact with later particles. Since this interaction may have negative effects, it is studied to determine its magnitude, and to determine any actions that may be taken to mitigate it.

Beam Dynamics

Due to the high velocity of the particles, and the resulting Lorentz force for magnetic fields, adjustments to the beam direction are mainly controlled by magnetostatic fields that deflect particles. In most accelerator concepts (excluding compact structures like the cyclotron or betatron), these are applied by dedicated electromagnets with different properties and functions. An important step in the development of these types of accelerators was the understanding of strong focusing. Dipole magnets are used to guide the beam through the structure, while quadrupole magnets are used for beam focusing, and sextupole magnets are used for correction of dispersion effects.

A particle on the exact design trajectory (or design *orbit*) of the accelerator only experiences dipole field components, while particles with transverse position deviation are re-focused to the design orbit. For preliminary calculations, neglecting all fields components higher than quadrupolar, an inhomogenic Hill differential equation

$$\frac{d^2}{ds^2}x(s)+k(s)x(s)=\frac{1}{R}\frac{\Delta p}{p}$$

can be used as an approximation, with

a non-constant focusing force $k(s)$, including strong focusing and weak focusing effects

the relative deviation from the design beam impulse $\Delta p / p$

the trajectory radius of curvature R, and

the design path length s,

thus identifying the system as a parametric oscillator. Beam parameters for the accelerator can then be calculated using Ray transfer matrix analysis; e.g., a quadrupolar field is analogous to a lens in geometrical optics, having similar properties regarding beam focusing (but obeying Earnshaw's theorem).

The general equations of motion originate from relativistic Hamiltonian mechanics, in almost all cases using the Paraxial approximation. Even in the cases of strongly non-linear magnetic fields, and without the paraxial approximation, a Lie transform may be used to construct an integrator with a high degree of accuracy.

Modeling Codes

There are many different software packages available for modeling the different aspects of accelerator physics. One must model the elements that create the electric and magnetic fields, and then one must model the charged particle evolution within those fields. A popular code for beam dynamics, designed by CERN is MAD, or Methodical Accelerator Design.

Beam Diagnostics

A vital component of any accelerator are the diagnostic devices that allow various properties of the particle bunches to be measured.

A typical machine may use many different types of measurement device in order to measure different properties. These include (but are not limited to) Beam Position Monitors (BPMs) to measure the position of the bunch, screens (fluorescent screens, Optical Transition Radiation (OTR) devices) to image the profile of the bunch, wire-scanners to measure its cross-section, and toroids or ICTs to measure the bunch charge (i.e., the number of particles per bunch).

While many of these devices rely on well understood technology, designing a device capable of measuring a beam for a particular machine is a complex task requiring much expertise. Not only is a full understanding of the physics of the operation of the device necessary, but it is also necessary to ensure that the device is capable of measuring the expected parameters of the machine under consideration.

Success of the full range of beam diagnostics often underpins the success of the machine as a whole.

Machine Tolerances

Errors in the alignment of components, field strength, etc., are inevitable in machines of this scale, so it is important to consider the tolerances under which a machine may operate.

Engineers will provide the physicists with expected tolerances for the alignment and manufacture of each component to allow full physics simulations of the expected behaviour of the machine under these conditions. In many cases it will be found that the performance is degraded to an unacceptable level, requiring either re-engineering of the components, or the invention of algorithms that allow the machine performance to be 'tuned' back to the design level.

This may require many simulations of different error conditions in order to determine the relative success of each tuning algorithm, and to allow recommendations for the collection of algorithms to be deployed on the real machine.

Particle Physics Experiments

Particle physics experiments briefly discusses a number of past, present, and proposed experiments with particle accelerators, throughout the world. In addition, some important accelerator interactions are discussed. Also, some notable systems components are discussed, named by project.

AEGIS (Particle Physics)

AEGIS is a proposed experiment to be set up at the Antiproton Decelerator at CERN. In addition, *AEGIS* is an acronym for: Antimatter Experiment: Gravity, Interferometry, Spectroscopy)

The proposed experiment:

It would attempt to determine if gravity affects antimatter in the same way it affects matter by testing its effect on an antihydrogen beam. By sending a stream of antihydrogen through a series of diffraction gratings, the pattern of light and dark patterns would allegedly enable the position of the beam to be pinpointed with up to 1% accuracy.

Athena

ATHENA was an antimatter research project that took place at the AD Ring at CERN. In 2005 ATHENA was disbanded and many of the former members became the ALPHA Collaboration. In August 2002, it was the first experiment to produce 50,000 low-energy antihydrogen atoms, as reported in the journal Nature.

For antihydrogen to be created, antiprotons and positrons must first be prepared. Once the antihydrogen is created, a high-resolution detector is needed to confirm that the antihydrogen was created, as well as to look at the spectrum of the antihydrogen in order to compare it to "normal" hydrogen.

The antiprotons are obtained from CERN's Antiproton Decelerator while the positrons are obtained from a positron accumulator. The antiparticles are then led into a recombination trap to create antihydrogen. The trap is surrounded by the ATHENA detector, which detects the annihilation of the antiprotons as well as the positrons.

The ATHENA Collaboration comprised the following institutions

*   University of Aarhus, Denmark

*   University of Brescia, Italy

*   CERN

*   University of Genoa, Italy

*   University of Pavia, Italy

*   RIKEN, Japan

*   Federal University of Rio de Janeiro, Brazil

*   University of Wales Swansea, United Kingdom

*   University of Tokyo, Japan

*   University of Zurich, Switzerland

*   National Institute for Nuclear Physics, Italy

ARGUS (Experiment)

The ARGUS experiment was a particle physics experiment that ran at the electron-positron collider ring *DORIS II* at DESY. It is the first experiment that observed the mixing of the B mesons (in 1987)

The ARGUS detector was a hermetic detector with 90% coverage of the full solid angle. It had drift chambers, a time-of-flight system, an electromagnetic calorimeter and a muon chamber system.

In physics, the ARGUS distribution, named after this experiment, is the probability distribution of the reconstructed invariant mass of a decayed particle candidate in continuum background. Its probability density function (not normalized) is:

$$f(x) = x \cdot \sqrt{1 - \left(\frac{x}{c}\right)^2} \, \exp\left\{ -\chi \cdot \left[1 - \left(\frac{x}{c}\right)^2 \right] \right\} \text{ for } x > 0.$$

Sometimes a more general form is used to describe a more peaking-like distribution:

$$f(x) = x \cdot \left[1 - \left(\frac{x}{c} \right)^2 \right]^p \exp \left\{ -\chi \cdot \left(1 - \left(\frac{x}{c} \right)^2 \right) \right\}$$

Here parameters c, χ, p represent the cutoff, curvature, and power ($p = 0.5$ gives a regular ARGUS) respectively.

ATRAP

The ATRAP collaboration at CERN developed out of TRAP, a collaboration whose members pioneered cold antiprotons, cold positrons, and first made the ingredients of cold antihydrogen to interact. ATRAP members also pioneered accurate hydrogen spectroscopy and first observed hot antihydrogen atoms. The collaboration includes investigators from Harvard, the University of Bonn, the Max Planck Institute for Quantum Optics, the University of Amsterdam, York University, Seoul National University, NIST, Forschungszentrum Jülich.

Belle Experiment

The Belle experiment is a particle physics experiment conducted by the Belle Collaboration, an international collaboration of more than 400 physicists and engineers investigating CP-violation effects at the High Energy Accelerator Research Organisation (KEK) in Tsukuba, Ibaraki Prefecture, Japan.

Systems Components

ASTRID Particle Storage Ring

ASTRID is a particle storage ring at the University of Aarhus, Århus, Denmark. It is located in the lower levels of the University of Aarhus Department of Physics and Astronomy.

Its construction was announced on 18 September 1987. By 1998, it had been improved several times, notably increasing its maximum operation time to 30–35 hours. In December 2008, a contract was awarded to design and build ASTRID 2, which will be built adjacent to ASTRID. ASTRID will be used to "top up" the new ring, allowing ASTRID 2 to operate nearly continuously.

ASTRID 2 Particle Storage Ring

ASTRID 2 will be a 46-meter particle storage ring at the University of Aarhus, Århus, Denmark. The contract to build the ring was awarded in December, 2008, and plans are expected to be complete by the end of 2009. It will be built in the lower levels of the

University of Aarhus Department of Physics and Astronomy, adjacent to the existing ASTRID particle storage ring. Rather than having an electron beam which decays over time, it will be continually "topped up" by a feed from ASTRID, allowing nearly constant current. It will generate synchrotron radiation to provide a tunable beam of light, expected to be of "remarkable" quality, with wavelengths from the ultraviolet through x-rays.

Anti-proton Decelerator

The antiproton decelerator (AD) is a storage ring at the CERN laboratory in Geneva. The decelerated antiprotons are ejected to one of the connected experiments.

Current Experiments

Expt.	Acronym	Full name
AD-2	ATRAP	Anti-hydrogen Trap Collaboration
AD-3	ASACUSA	Atomic Spectroscopy And Collisions Using Slow Anti-protons
AD-4	ACE	Anti-proton Cell Experiment
AD-5	ALPHA	Anti-hydrogen Laser Physics Apparatus
AD-6	AEGIS	Anti-hydrogen Experiment: Gravity, Interferometry, Spectroscopy

Former Experiments:

Expt.	Acronym	Full name
AD-1	ATHENA	ApparaTus for High-precision Experiments on Neutral Antimatter

Accelerator Interaction Overview

Absorber

In high energy physics experiments, an absorber is a block of material used to absorb some of the energy of an incident particle. Absorbers can be made of a variety of materials, depending on the purpose; lead and liquid hydrogen are common choices.

Most absorbers are used as part of a detector.

A more recent use for absorbers is for ionization cooling, as in the International Muon Ionization Cooling Experiment.

In solar power, the most important part of the collector takes up the heat of the solar radiation through a medium (water + antifreeze). This is heated and circulates between the collector and the storage tank. A high degree of efficiency is achieved by using black absorbers or, even better, through selective coating.

In sunscreen, ingredients which absorb UVA/UVB rays, such as avobenzone and octyl methoxycinnamate, are known as absorbers. They are contrasted with physical "blockers" of UV radiation such as titanium dioxide and zinc oxide.

Accelerator Physics

Accelerator physics is an interdisciplinary topic of applied physics, commonly defined by the intent of designing, building and operating particle accelerators.

The experiments conducted with particle accelerators are not regarded as part of accelerator physics, but belong (according to the objectives of the experiments) to e.g. particle physics, nuclear physics, condensed matter physics or materials physics. The types of experiments done at a particular accelerator facility are determined by characteristics of the generated particle beam such as average energy, particle type, intensity, and dimensions.

Event Reconstruction

In a particle detector experiment, event reconstruction is the process of interpreting the electronic signals produced by the detector to determine the original particles that passed through, their momenta, directions, and the primary vertex of the event. Thus the initial physical process that occurred at the interaction point of the particle accelerator, whose study is the ultimate goal of the experiment, can be determined. The total event reconstruction is rarely possible (and rarely necessary); usually, only some part of the data described above is obtained and processed.

References

- Humphries, Stanley (1986). Principles of Charged Particle Acceleration. Wiley-Interscience. p. 4. ISBN 978-0471878780.

- "Atom smasher". American Heritage Science Dictionary. Houghton Mifflin Harcourt. 2005. p. 49. ISBN 978-0-618-45504-1.

- Hamm, Robert W.; Hamm, Marianne E. (2012). Industrial Accelerators and Their Applications. World Scientific. ISBN 978-981-4307-04-8.

- Humphries, Stanley (1986). "Linear Induction Accelerators". Principles of Charged Particle Acceleration. Wiley-Interscience. pp. 283–325. ISBN 978-0471878780.

- Chao, A. W.; Mess, K. H.; Tigner, M.; et al., eds. (2013). Handbook of Accelerator Physics and Engineering (2nd ed.). World Scientific. ISBN 978-981-4417-17-4.

- Humphries, Stanley (1986). "Betatrons". Principles of Charged Particle Acceleration. Wiley-Interscience. p. 326ff. ISBN 978-0471878780.

- Schopper, Herwig F. (1993). Advances of accelerator physics and technologies. World Scientific. ISBN 981-02-0957-6. Retrieved March 9, 2012.

- Wiedemann, Helmut (1995). Particle accelerator physics 2. Nonlinear and higher-order beam dynamics. Springer. ISBN 0-387-57564-2. Retrieved March 9, 2012.

- Chao, Alex W.; Tigner, Maury, eds. (2013). Handbook of accelerator physics and engineering

(2nd ed.). World Scientific. ISBN 978-981-4417-17-4.

- Witman, Sarah. "Ten things you might not know about particle accelerators". Symmetry Magazine. Fermi National Accelerator Laboratory. Retrieved 21 April 2014.

- Peralta, E. A.; et al. "Demonstration of electron acceleration in a laser-driven dielectric microstructure". Retrieved 2014-05-01.

- Nielsen, J. S.; Møller, S. P. (1998). "New Developments at the ASTRID storage ring" (PDF). Proceedings from 6th European Particle Accelerator Conference. Retrieved 2012-04-10.

7

Evolution of Particle Physics

The idea of particles and matter has existed in natural philosophy since 6th century BC. Particles have been discovered and researched; these particles are studied under high energies. This text helps the reader in developing an understanding related to the history of particle physics.

The idea that matter consists of smaller particles and that there exists a limited number of sorts of primary, smallest particles in nature has existed in natural philosophy at least since the 6th century BC. Such ideas gained physical credibility beginning in the 19th century, but the concept of "elementary particle" underwent some changes in its meaning: notably, modern physics no longer deems elementary particles indestructible. Even elementary particles can decay or collide destructively; they can cease to exist and create (other) particles in result.

A Crookes tube with a magnetic deflector

Increasingly small particles have been discovered and researched: they include molecules, which are constructed of atoms, that in turn consist of subatomic particles, namely atomic nuclei and electrons. Many more types of subatomic particles have been found. Most such particles (but not electrons) were eventually found to be composed of even smaller particles such as quarks. Particle physics studies these smallest particles and their behaviour under high energies, whereas nuclear physics studies atomic nuclei and their (immediate) constituents: protons and neutrons.

Early Development

The idea that all matter is composed of elementary particles dates to at least the 6th century BC. The philosophical doctrine of atomism and the nature of elementary particles were studied by ancient Greek philosophers such as Leucippus, Democritus, and Epicurus; ancient Indian philosophers such as Kanada, Dignāga, and Dharmakirti; Muslim scientists such as Ibn al-Haytham, Ibn Sina, and Mohammad al-Ghazali; and in early modern Europe by physicists such as Pierre Gassendi, Robert Boyle, and Isaac Newton. The particle theory of light was also proposed by Ibn al-Haytham, Ibn Sina, Gassendi, and Newton.

Those early ideas were founded through abstract, philosophical reasoning rather than experimentation and empirical observation and represented only one line of thought among many. In contrast, certain ideas of Gottfried Wilhelm Leibniz contradict to almost everything known in modern physics.

In the 19th century, John Dalton, through his work on stoichiometry, concluded that each element of nature was composed of a single, unique type of particle. Dalton and his contemporaries believed those were the fundamental particles of nature and thus named them atoms, after the Greek word *atomos*, meaning "indivisible" or "uncut".

From Atoms to Nucleons

First Subatomic Particles

However, near the end of 19th century, physicists discovered that Dalton's atoms are not, in fact, the fundamental particles of nature, but conglomerates of even smaller particles. Electron was discovered between 1879 and 1897 in works of William Crookes, Arthur Schuster, J. J. Thomson, and other physicists; its charge was carefully measured by Robert Andrews Millikan and Harvey Fletcher in their oil drop experiment of 1909. Physicists theorized that negatively charged electrons are constituent part of "atoms", along with some (yet unknown) positively charged substance, and it was later confirmed. Electron became the first elementary, truly fundamental particle discovered.

Studies of the "radioactivity", that soon revealed the phenomenon of radioactive decay, provided another argument against considering chemical elements as fundamental nature's elements. Despite these discoveries, the term *atom* stuck to Dalton's (chemical) atoms and now denotes the smallest particle of a chemical element, not something really indivisible.

Researching Particles' Interaction

Early 20th-century physicists knew only two fundamental forces: electromagnetism and gravitation, where the latter could not explain the structure of atoms. So, it was obvious to assume that unknown positively charged substance attracts electrons by Coulomb force.

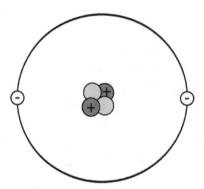

In 1909 Ernest Rutherford and Thomas Royds demonstrated that an alpha particle combines with two electrons and forms a helium atom. In modern terms, alpha particles are doubly ionized helium (more precisely, 4He) atoms. Speculation about the structure of atoms was severely constrained by Rutherford's 1907 gold foil experiment, showing that the atom is mainly empty space, with almost all its mass concentrated in a tiny atomic nucleus.

Inside the Atom

By 1914, experiments by Ernest Rutherford, Henry Moseley, James Franck and Gustav Hertz had largely established the structure of an atom as a dense nucleus of positive charge surrounded by lower-mass electrons. These discoveries shed a light to the nature of radioactive decay and other forms of transmutation of elements, as well as of elements themselves. It appeared that atomic number is nothing else than (positive) electric charge of the atomic nucleus of a particular atom. Chemical transformations, governed by electromagnetic interactions, do not change nuclei – that's why elements are chemically indestructible. But when the nucleus change its charge and/or mass (by emitting or capturing a particle), the atom can become the one of another element. Special relativity explained how the *mass defect* is related to the energy produced or consumed in reactions. The branch of physics that studies transformations and the structure of nuclei is now called nuclear physics, contrasted to atomic physics that studies the structure and properties of atoms ignoring most nuclear aspects. The development in the nascent quantum physics, such as Bohr model, led to the understanding of chemistry in terms of the arrangement of electrons in the mostly empty volume of atoms.

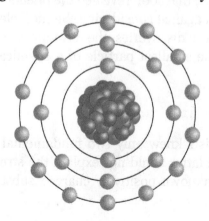

In 1918, Rutherford confirmed that the hydrogen nucleus was a particle with a positive charge, which he named the proton. By then, Frederick Soddy's researches of radioactive elements, and experiments of J. J. Thomson and F.W. Aston conclusively demonstrated existence of isotopes, whose nuclei have different masses in spite of identical atomic numbers. It prompted Rutherford to conjecture that all nuclei other than hydrogen contain chargeless particles, which he named the neutron. Evidences that atomic nuclei consist of some smaller particles (now called *nucleons*) grew; it became obvious that, while protons repulse each other electrostatically, nucleons attract each other by some new force (nuclear force). It culminated in proofs of nuclear fission in 1939 by Lise Meitner (based on experiments by Otto Hahn), and nuclear fusion by Hans Bethe in that same year. Those discoveries gave rise to an active industry of generating one atom from another, even rendering possible (although it will probably never be profitable) the transmutation of lead into gold; and, those same discoveries also led to the development of nuclear weapons.

Revelations of Quantum Mechanics

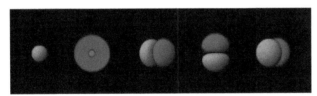

Atomic orbitals of Period 2 elements:
1s   2s   2p (3 items).

All complete subshells (including 2p) are inherently spherically symmetric, but it is convenient to assign to "distinct" p-electrons these two-lobed shapes.

Further understanding of atomic and nuclear structures became impossible without improving the knowledge about the essence of particles. Experiments and improved theories (such as Erwin Schrödinger's "electron waves") gradually revealed that there is no fundamental difference between particles and waves. For example, electromagnetic waves were reformulated in terms of particles called *photons*. It also revealed that physical objects do not change their parameters, such as total energy, position and momentum, as continuous functions of time, as it was thought of in classical physics: see atomic electron transition for example.

Another crucial discovery was identical particles or, more generally, quantum particle statistics. It was established that all electrons are identical: although two or more electrons can exist simultaneously that have different parameters, but they do not keep separate, distinguishable histories. This also applies to protons, neutrons, and (with certain differences) to photons as well. It suggested that there is a limited number of sorts of smallest particles in the universe.

The spin–statistics theorem established that any particle in our spacetime may be either a boson (that means its statistics is Bose–Einstein) or a fermion (that means its statistics is Fermi–Dirac). It was later found that all fundamental bosons transmit forces, like the photon that transmits light. Some of non-fundamental bosons (namely, mesons) also may transmit forces, although non-fundamental ones. Fermions are particles "like electrons and nucleons" and generally comprise the matter. Note that any subatomic or atomic particle composed of even *total* number of fermions (such as protons, neutrons, and electrons) is a boson, so a boson is not necessarily a force transmitter and perfectly can be an ordinary material particle.

The spin is the quantity that distinguishes bosons and fermions. Practically it appears as an intrinsic angular momentum of a particle, that is unrelated to its motion but is linked with some other features like a magnetic dipole. Theoretically it is explained from different types representations of symmetry groups, namely tensor representations (including vectors and scalars) for bosons with their integer (in $\hbar$) spins, and spinor representations for fermions with their half-integer spins.

This culminated in the formulation of ideas of a quantum field theory. The first (and the only mathematically complete) of these theories, quantum electrodynamics, allowed to explain thoroughly the structure of atoms, including the Periodic Table and atomic spectra. Ideas of quantum mechanics and quantum field theory were applied to nuclear physics too. For example, α decay was explained as a quantum tunneling through nuclear potential, nucleons' fermionic statistics explained the nucleon pairing, and Hideki Yukawa proposed certain virtual particles (now knows as π-mesons) as an explanation of the nuclear force.

Inventory

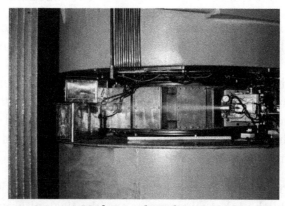

Modern nuclear physics

Development of nuclear models (such as the liquid-drop model and nuclear shell model) made prediction of properties of nuclides possible. No existing model of nucleon–nucleon interaction can *analytically* compute something more complex than 4He based on principles of quantum mechanics, though (note that complete computation of electron shells in atoms is also impossible yet).

The most developed branch of nuclear physics in 1940s was studies related to nuclear fission due to its military significance. The main focus of fission-related problems is interaction of atomic nuclei with neutrons: a process that occurs in a fission bomb and a nuclear fission reactor. It gradually drifted away from the rest of subatomic physics and virtually became the nuclear engineering. First synthesised transuranium elements were also obtained in this context, through neutron capture and subsequent β^- decay.

The elements beyond fermium cannot be produced in this way. To make a nuclide with more than 100 protons per nucleus one has to use an inventory and methods of particle physics, namely to accelerate and collide atomic nuclei. Production of progressively heavier synthetic elements continued into 21st century as a branch of nuclear physics, but only for scientific purposes.

The third important stream in nuclear physics are researches related to nuclear fusion. This is related to thermonuclear weapons (and conceived peaceful thermonuclear energy), as well as to astrophysical researches, such as stellar nucleosynthesis and Big Bang nucleosynthesis.

Physics Goes to High Energies

Strange Particles and Mysteries of the Weak Interaction

In the 1950s, with development of particle accelerators and studies of cosmic rays, inelastic scattering experiments on protons (and other atomic nuclei) with energies about hundreds of MeVs became affordable. They created some short-lived resonance "particles", but also *hyperons* and *K-mesons* with unusually long lifetime. The cause of the latter was found in a new quasi-conserved quantity, named *strangeness*, that is conserved in all circumstances except for the weak interaction. The strangeness of heavy particles and the μ-lepton were first two signs of what is now known as the second generation of fundamental particles.

The weak interaction revealed soon yet another mystery. In 1957 it was found that it does not conserve parity. In other words, the mirror symmetry was disproved as a fundamental symmetry law.

Throughout the 1950s and 1960s, improvements in particle accelerators and particle detectors led to a bewildering variety of particles found in high-energy experiments. The term *elementary particle* came to refer to dozens of particles, most of them unstable. It prompted Wolfgang Pauli's remark: "Had I foreseen this, I would have gone into botany". The entire collection was nicknamed the "particle zoo". It became evident that some smaller constituents, yet invisible, form mesons and baryons that counted most of then-known particles.

Deeper Constituents of Matter

The interaction of these particles by scattering and decay provided a key to new fundamental quantum theories. Murray Gell-Mann and Yuval Ne'eman brought some order to mesons and baryons, the most numerous classes of particles, by classifying them according to certain qualities. It began with what Gell-Mann referred to as the "Eightfold Way", but proceeding into several different "octets" and "decuplets" which could predict new particles, most famously the $\Omega-$, which was detected at Brookhaven National Laboratory in 1964, and which gave rise to the *quark* model of hadron composition. While the quark model at first seemed inadequate to describe s trong n uclear f orces, allowing the temporary rise of competing theories such as the S-matrix theory, the establishment of quantum chromodynamics in the 1970s finalized a set of fundamental and exchange particles (Kragh 1999). It postulated the fundamental strong interaction, experienced by quarks and mediated by gluons. These particles were proposed as a building material for hadrons. This theory is unusual because individual (free) quarks cannot be observed (color confinement), unlike the situation with composite atoms where electrons and nuclei can be isolated by transferring ionization energy to the atom.

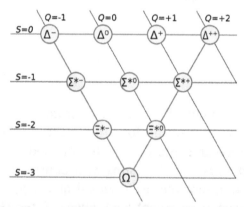

Classification of spin-3/2 baryons known in 1960s

Then, the old, broad denotation of the term *elementary particle* was deprecated and a replacement term *subatomic particle* covered all the "zoo", with its hyponym "hadron" referring to composite particles directly explained by the quark model. The designation of an "elementary" (or "fundamental") particle was reserved for leptons, quarks, their antiparticles, and quanta of fundamental interactions only.

Quarks, Leptons, and Four Fundamental Forces

Because the quantum field theory postulates no difference between parti-cles and interactions, classification of elementary particles allowed also to classify interactions and fields.

Now a large number of particles and (non-fundamental) interactions is explained as combinations of a (relatively) small number of fundamental substances, thought to be fundamental interactions (incarnated in fundamental bosons), quarks (including antiparticles), and leptons (including antiparticles). As the theory distinguished *several* fundamental interactions, it became possible to see which elementary particles participate in which interaction. Namely:

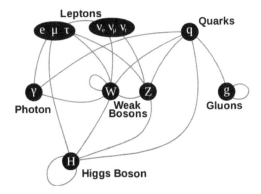

- All particles participate in gravitation.

- All charged elementary particles participate in electromagnetic interaction.

  o As a consequence, neutron participate in it with its magnetic dipole in spite of zero electric charge. This is because it is composed of *charged* quarks whose charges sum to zero.

- All fermions participate in the weak interaction.

- Quarks participate in the strong interaction, along gluons (its own quanta), but not leptons nor any fundamental bosons other than gluons.

The next step was a reduction in number of fundamental interactions, envisaged by early 20th century physicists as the "united field theory". The first successful modern unified theory was the electroweak theory, developed by Abdus Salam, Steven Weinberg and, subsequently, Sheldon Glashow. This development culminated in the completion of the theory called the Standard Model in the 1970s, that included also the strong interaction, thus covering three fundamental forces. After the discovery, made at CERN, of the existence of neutral weak currents, mediated by the Z boson foreseen in the standard model, the physicists Salam, Glashow and Weinberg received the 1979 Nobel Prize in Physics for their electroweak theory. The discovery of the weak gauge bosons (quanta of the weak interaction) through the 1980s, and the verification of their properties through the 1990s is considered to be an age of consolidation in particle physics.

While accelerators have confirmed most aspects of the Standard Model by detecting expected particle interactions at various collision energies, no theory reconciling general relativity with the Standard Model has yet been found, although supersymmetry and string theory were believed by many theorists to be a promising avenue forward. The Large Hadron Collider, however, which began operating in 2008, has failed to find any evidence whatsoever that is supportive of supersymmetry and string theory, and appears unlikely to do so, meaning "the current situation in fundamental theory is one of a serious lack of any new ideas at all." This state of affairs should not be viewed as a crisis in physics, but rather, as David Gross has said, "the kind of acceptable scientific confusion that discovery eventually transcends."

The fourth fundamental force, gravitation, is not yet integrated into particle physics in a consistent way.

Higgs Boson

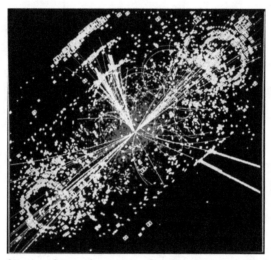

One possible signature of a Higgs boson from a simulated proton–proton collision. It decays almost immediately into two jets of hadrons and two electrons, visible as lines.

As of 2011, the Higgs boson, the quantum of a field that is thought to provide particles with rest masses, remained the only particle of the Standard Model to be verified. On July 4, 2012, physicists working at CERN's Large Hadron Collider announced that they had discovered a new subatomic particle greatly resembling the Higgs boson, a potential key to an understanding of why elementary particles have masses and indeed to the existence of diversity and life in the universe. Rolf-Dieter Heuer, the director general of CERN, said that it was too soon to know for sure whether it is an entirely new particle, which weighs in at 125 billion electron volts – one of the heaviest subatomic particles yet – or, indeed, the elusive particle predicted by the Standard Model, the theory that has ruled physics for the last half-century. It is unknown if this particle is an impostor, a single particle or even the first of many particles yet to be discovered. The latter possibilities are particularly exciting to physicists since they could point the way to new deeper ideas, beyond the Standard Model, about the nature of reality. For now, some physicists are calling it a "Higgslike" particle. Joe Incandela, of the University of California, Santa Barbara, said, "It's something that may, in the end, be one of the biggest observations of any new phenomena in our field in the last 30 or 40 years, going way back to the discovery of quarks, for example." The groups operating the large detectors in the collider said that the likelihood that their signal was a result of a chance fluctuation was less than one chance in 3.5 million, so-called "five sigma," which is the gold standard in physics for a discovery. Michael Turner, a cosmologist at the University of Chicago and the chairman of the physics center board, said

This is a big moment for particle physics and a crossroads — will this be the high water mark or will it be the first of many discoveries that point us toward solving the really big questions that we have posed?

—Michael Turner, University of Chicago

Confirmation of the Higgs boson or something very much like it would constitute a rendezvous with destiny for a generation of physicists who have believed the boson existed for half a century without ever seeing it. Further, it affirms a grand view of a universe ruled by simple and elegant and symmetrical laws, but in which everything interesting in it being a result of flaws or breaks in that symmetry. According to the Standard Model, the Higgs boson is the only visible and particular manifestation of an invisible force field that permeates space and imbues elementary particles that would otherwise be massless with mass. Without this Higgs field, or something like it, physicists say all the elementary forms of matter would zoom around at the speed of light; there would be neither atoms nor life. The Higgs boson achieved a notoriety rare for abstract physics. To the eternal dismay of his colleagues, Leon Lederman, the former director of Fermilab, called it the "God particle" in his book of the same name, later quipping that he had wanted to call it "the goddamn particle". Professor Incandela also stated,

This boson is a very profound thing we have found. We're reaching into the fabric of the universe at a level we've never done before. We've kind of completed one particle's story [...] We're on the frontier now, on the edge of a new exploration. This could be the only part of the story that's left, or we could open a whole new realm of discovery.

—Joe Incandela, University of California

In quantum theory, which is the language of particle physicists, elementary particles are divided into two rough categories: fermions, which are bits of matter like electrons, and bosons, which are bits of energy and can transmit forces, like the photon that transmits light. Dr. Peter Higgs was one of six physicists, working in three independent groups, who in 1964 invented the notion of the cosmic molasses, or Higgs field. The others were Tom Kibble of Imperial College, London; Carl Hagen of the University of Rochester; Gerald Guralnik of Brown University; and François Englert and Robert Brout, both of Université Libre de Bruxelles. One implication of their theory was that this Higgs field, normally invisible and, of course, odorless, would produce its own quantum particle if hit hard enough, by the right amount of energy. The particle would be fragile and fall apart within a millionth of a second in a dozen different ways depending upon its own mass. Unfortunately, the theory did not say how much this particle should weigh, which is what made it so difficult to find. The particle eluded researchers at a succession of particle accelerators, including the Large Electron–Positron Collider at CERN, which closed down in 2000, and the Tevatron at the Fermi National Accelerator Laboratory, or Fermilab, in Batavia, Ill., which shut down in 2011.

Further experiments continued and in March 2013 it was tentatively confirmed that the newly discovered particle was a Higgs Boson.

Although they have never been seen, Higgslike fields play an important role in theories of the universe and in string theory. Under certain conditions, according to the strange accounting of Einsteinian physics, they can become suffused with energy that exerts an antigravitational force. Such fields have been proposed as the source of an enormous burst of expansion, known as inflation, early in the universe and, possibly, as the secret of the dark energy that now seems to be speeding up the expansion of the universe.

Further Theoretical Development

Modern theoretical development includes refining of the Standard Model, researching in its foundations such as the Yang–Mills theory, and researches in computational methods such as the lattice QCD.

A long-standing problem is quantum gravitation. No solution that is useful for particle physics has been achieved.

Further Experimental Development

There are researches about quark–gluon plasma, a new (hypothetical) state of matter. There are also some recent experimental evidences that tetraquarks and glueballs exist.

The proton decay is not observed (or, generally, non-conservation of the baryon number), but predicted by the Standard Model, so there are searches for it.

References

- Smirnov, B.M. (2003). Physics of Atoms and Ions. Springer. pp. 14–21. ISBN 0-387-95550-X.
- Rincon, Paul (2012-07-04). "BBC News - Higgs boson-like particle discovery claimed at LHC". Bbc.co.uk. Retrieved 2013-04-20.

Permissions

Index

www.ingramcontent.com/pod-product-compliance
Lightning Source LLC
Jackson TN
JSHW050758150125
77033JS00052B/178